INTERPRETING NATURE

Human society has constructed many varied notions of the environment. Scientific information about the environment is often seen as the only worthwhile knowledge. This ignores the complexities created by interaction between people and the environment. Idealist thinking argues that everything we know is based on a construct of our minds and that all is possible. Can both be correct and true?

Interpreting Nature explores the position of humanity in the environment from the principle that the models we construct are imperfect and can only be provisional. Concentrating on ideas rather than instrumentality, the book discusses the environment in an historical, cultural context. Having examined the way in which the natural sciences have interrogated nature, the types of data produced and what they mean to us, the author introduces the reader to thinking about the environment in philosophy and ethics, the social sciences and the arts, and analyses their role in the formation of environmental cognition. Extensive bibliographies allow the opportunity to follow up any of the matters discussed.

I.G. Simmons is Professor of Geography at the University of Durham.

D1125302

INTERPRETING NATURE

Cultural constructions
of the environment

I. G. Simmons

London and New York

First published 1993
by Routledge
11 New Fetter Lane

Simultaneously published in the USA and Canada
by Routledge
29 West 35th Street, New York, NY 10001

© 1993 Ian Simmons

Typeset in Garamond by
J&L Composition Ltd, Filey, North Yorkshire
Printed and bound in Great Britain by
Biddles Ltd, Guildford and King's Lynn

British Library Cataloguing in Publication Data
A catalogue reference for this book is available from the British Library.

Library of Congress Cataloging in Publication Data
has been applied for

ISBN 0–415–09705–3
0–415–09706–1

CONTENTS

CONTENTS

PLATES

FIGURES

TABLES

PREFACE

When I finished the text of *Changing the Face of the Earth* (Blackwell, Oxford, 1989), I felt that the story was incomplete. Obviously, because in trying to cover so much space and time, quite a lot of material had been left out. Less obvious, though I alluded to it in that book's Preface, was the fact that it was a naively realistic account, which simply took 'facts' at their face value as recorded by historians, palaeoanthropologists, historical geographers *et al*. There was no real place in that book for further explorations of the cultural filters of many kinds through which that information had come, and there never was the intention that it should essay that task. (The explanation of that in the book did not, however, prevent one reviewer from lamenting quite sharply the absence of such an approach.) But by then I was convinced that there was more to environmental history than facts and there was indeed a whole series of cultural frameworks by means of which these facts came to be so regarded.

Teaching an optional course to third year undergraduates on Environmental Thought had also convinced me that most students who considered today's concern about the environment did so in the context of a linear model. Namely, that the facts were provided by the natural sciences and everything else was simply a matter of how we responded to those actualities. It took me a lot of effort to persuade them even to consider that there might be even deeper cultural frameworks in which science, for example, might be embedded. (I might add that it was clearly an idea foreign to some of the external examiners who read their examination papers, too.) I was also even more keenly aware that the undergraduates knew very little about intellectual history and hence about the roots of the environmental thought that goes on today.

So bringing these two realisations together made me undertake a lot of reading and a few other (often pleasant) tasks as well, like seeing a number of paintings at first hand rather than in reproductions, and trying to put together in a book the main lineaments of environmental thought in the West today in a form which undergraduates in subjects like Environmental Science and Geography can understand. That is an important beginning point: I have tried as far as I can to start where they start, and that is, in general, with a notion that there are different disciplines with different approaches to the Earth and its inhabitants and their linking systems. So I have simply started there, with an elementary set of categories, and gone onward to explore the range (and indeed richness) of the contents of each category for our approach to that entity which we label 'environment'. Some basic follow-up reading is suggested for each chapter, and to show fellow academics where I have been (and what I have missed) there are some fairly extensive endnotes; these also explain some of the terms used and so form a sort of parallel text, with some comments and annotations which I judged would break up the progress of the main themes. More advanced readers will, I am sure, get very frustrated by the fact that the accounts seem to stop just when the story is getting satisfactorily complicated. This is because the book is specifically an introduction to the matters within its remit, and they will have to follow them up if they want more: that means reading the specialists in the different fields of scholarship which are united by their dealing with 'environment'. Since the book is aimed primarily at geographers, it also serves notice on them that I think they have to break out of some of their traditional notions of what Geography can deal with if they are to have anything to say about 'environment'. They may not in general wish to, of course: in the 1965–72 period of 'the environmental crisis' their attention was largely directed elsewhere.

I have not attempted to write a history of environmental ideas. There are already one or two useful books on that scene (e.g. David Pepper's *The Roots of Modern Environmentalism*, Croom Helm, 1984) and there seems no point in repeating their contents. So history in this book acts only as an introductory background. We badly need, though, equivalents of Clarence Glacken's *Traces on the Rhodian Shore* (University of California Press, 1967) for the nineteenth and twentieth centuries in the West, and comparable work for East and South Asia, Africa and Latin America. So this is a book dealing with the recent past and the immediate future; it attempts no grand statements about origins nor prescriptions about what to do next. It is not, therefore, an environmentalist tract, to be spectrographically analysed for its shade of viridity. If it has a specifically didactic purpose beyond that of trying to interest students in matters of ideas rather than of instrumentality, then it is to suggest that the norm is complexity and richness rather than binary opposites of black and white or of deep green and flaming red. Indeed, the working title for the book has in my mind always been 'Nine-and-Sixty Ways', from Rudyard Kipling's *In the Neolithic Age*:

> There are nine and sixty ways of constructing tribal lays
> And – every – single – one – of – them – is – right

but successive editors have choked into their *fettucine* (if in London) or cheese toasties (if in Durham) and so we have settled upon a more immediately self-explanatory title.

PREFACE

In writing this book, I was fortunate to have had a period of leave from the University of Durham in 1987–8 in which to get started and to lay the foundations of the text. Ever since then, it has been rewritten and incrementally augmented at intervals, some of which were relatively long as I tried to find out whether a particular take-over in the publishing world had or had not included the contract for this volume, and I am grateful to Tristan Palmer for his encouraging confirmation that this imprint was indeed going ahead with the project.

As always I need to acknowledge the help of a number of people. Even though I have been Chairman of my department during 1990–2, my colleagues agreed to an administrative structure that would not totally preclude me doing some academic work. It was not their fault that these were the years of the Audit (two of these), Departmental and Faculty Plans, a Research Assessment Exercise and quite a lot of financial juggling. So this is what (*inter alia*) I was doing on Thursdays: my grateful thanks to everyone who put up with my non-availability. Leslie Yeung was especially effective in deterring would-be telephone callers on such days. Various parts of the text have been tried out on seminar and other groups here and there and I am grateful for feedback on these occasions. The line-work was created by the department's cartographic staff, and I thank them for their skill. The later versions of the text were corrected and amended in minute detail by Kathy Wood, whom I also thank, especially for her patience. The book would simply not have been possible without the help of the staff of the Inter-Library Loan desk of the University Library: thank you Anne, Fran and Gwynneth. In spite of everything that has happened on the wider scene recently, the University of Durham, and its Department of Geography, remain good places to be, and I am very grateful to those who strive to keep it so.

To break away from thirty years of science-based empirical research into this sort of material is a considerable risk and I doubt if I would have undertaken it without the enormous encouragement from Carol, my wife. She knows just how much. This support came in many forms but chiefly in stopping me working. You may think she should have intervened much sooner; I hope not.

I. G. Simmons
Durham, Summer 1992

1

INTRODUCTION

To begin at the beginning always sounds easy, so to talk of the interaction of humans and their environment appears a self-evident task. However if we examine more closely the terms 'human' and 'environment', then certain complexities emerge. The biological classification of present-day humans is not very difficult, for the genus has only the one species. Yet this particular species occupies not only an ecological world but a psychological one too. As well as its biophysical surroundings, it has an environment which we understand culturally. Hence, how we act towards the non-human is a consequence of our beliefs both about ourselves and what it is we are acting upon. An advertising leaflet for *New Scientist* in the late 1980s was dominated by a full-colour reproduction of the famous photograph of Earth from space, and the caption read, 'What on earth is going on *out there?*' (my italics). Also, human culture involves not just individuals but also groups of various sizes which do not always act as a simple sum of their numbers. For example, such collections may adopt patterns of behaviour which either increase or diminish their impact on their surroundings in ways which are not related simply to population totals.

To define environment is a job of altogether greater complexity. It seems easy at first if we accept the commonly used sense that it consists of all those material entities which exist on planet Earth but which are not human.[1] If we accept this view then we have immediately a framework in which to discuss 'the environment' but we are open to accusations of epistemological shallowness.[2] We accept a dualism of 'humans' and 'environment' which may not be true in reality since we ourselves are the only reference point: there is no totally detached outsider who can see the whole picture.[3] Further, it encourages a static picture of 'us' and 'it' which clearly pays little heed to the dynamics of the processes which connect living humans to other things. Lastly, not all cultures accept that such a dualism describes a real division.

Yet there is another view, that what we call 'environment' only exists in the human mind; it is an artefact of mental processes and therefore is quite inseparable from the human. In these terms, to talk of some separate reality

1

of phenomena like plants and the sea is nonsense, and particularly so if we recognise the problems of language. Language may, it seems, impose a reality rather than simply attach labels to what is 'there'. But to suggest that humans and the rest of the cosmos are one undifferentiated set of mind and matter is to run the risk of paralysis in the face of the fact that people obviously do behave as if non-human things and processes are qualitatively different from humans themselves. What then is the place of humankind in the universe? Many thinkers have had a crack at that question: especially relevant to our present theme is the **anthropic principle** which in its 'weak' form says that the conditions for the development of intelligent life in the universe seem to be limited. So we ought not to be surprised if our locality satisfies the conditions for our existence. In a strong form, it can be 'a staggering realisation that the composition of the universe is so exquisitely balanced that it makes life not only possible but perhaps necessary. The universe must have known we were coming'. Even stronger is the introduction from quantum physics that observers (and only humans are eligible for this status) are necessary to bring the universe into being. Against this, others have argued that the principle negates the possibility of chance having played any role in cosmic and organic evolution; and further that a great deal seems to have happened before the emergence of *Homo sapiens* and so there is no need for the principle at all. What such discussion shows us at this point is the range of argument over what seems at first a relatively simple question, to which we might have answered, 'we are the people at the top of the evolutionary ladder and our intelligence has allowed us to do some remarkable things'. So our discussions of environmental thinking and knowing are framed by discussions of a very basic and broad-ranging kind about the nature of humanity and of the universe.[4]

Such discussions bring home the realisation that simple, dictionary-like, definitions of 'environment' will not suffice for a close look at the relations of humanity and environment. Many users simply mean the biological and physico-chemical systems of this planet outside the bodies of humans, though not nowadays excluding the results of human actions in those systems. It can appear to change independently of human action or thought and also to limit the actions of our species. Thus there is a set of processes which enmeshes humans (as organisms created by cosmic processes) with the rest of the edifice, but we can designate ourselves as qualitatively different since we can create ourselves psychologically to a great extent and in so doing alter our understanding about what the rest of the world is like. So when we use the term with this fullest set of meanings, it is rich, diverse, and elusive.[5]

One immediate aim, therefore, is to find a way of discussing the theme of this work which will acknowledge that any framework is provisional, may well be culture-bound and indeed may be a poor reflection of what 'reality' is actually like. The biologist J. B. S. Haldane (1892–1964) said that not only was the world queerer than we supposed but was probably queerer than we *could* suppose. Put more formally, it seems as if there are any number of competing ontologies[6] to choose from, leading to:

> the realization that we cannot in principle distinguish between the constructed nature of our intelligible world and the 'independent' structure of the brute world[7]

turning our task in this volume into a ride on a carousel from which it is impossible to alight.

One way of stopping this roundabout, if only temporarily, is to develop the idea of a construction.[8] Here we accept that there is indeed a 'real' cosmos but that we are too limited to comprehend its true nature. We have too few data, for example, or our brains are not big enough to comprehend the complexity of all that we perceive, even with the help of books or computers.[9] In order therefore to reduce the mass of information to something which we can tell ourselves that we understand (and to make a living for the academic staff of institutions of Higher Education) and especially so that somebody can do something about it, we make constructions of various kinds. This book will be organised about the principle that there are different types of constructions of the world made by various groups of thinkers and doers; we shall accept from this moment onwards that they are all imperfect and can be only provisional. It will be a task for the end of the book to ask whether all the constructions have in them flaws that make them so imperfect that a different type of 'knowing' altogether is a preferable option.

TYPES OF CONSTRUCTIONS

We will accept for the first three chapters that what is beyond our own skin actually exists. But this 'environment' is largely what we make of it, with all the ambiguities inherent in the word 'make'. In the constructions which are in play today, there are however some fundamental differences in the way that external entity is understood. There are, for example, those constructions which are claimed to be **objective**,[10] i.e. the observer is a detached consciousness (like a clean slate or *tabula rasa*) and so all the phenomena and processes described are external to his/her mind before they are brought in, so to speak, for description and explanation. Public verifiability is therefore possible: measurements made of an object or a process will be the same whether the observers are in Ōsaka or Oxford. This is classically the realm of the natural scientist but social scientists may attempt to produce studies of humans and their societies using this methodology.

An apparently different type of knowledge is labelled **subjective**,[11] because it is internal to the observer and may thus vary according to that observer's

3

individual characteristics. This category of information is central to the humanities and the fine arts: it is not reckoned that everybody should create or respond to a novel or a sculpture in exactly the same way. It too extends to the social sciences: for some investigators, human experience in all its variety is the valid starting point. As an instance we might think of noise. Objectively, the sound of my son's record-player can be measured in decibels and the number obtained is directly related to the output of sound, i.e. the quantity of energy produced. A given number on the volume knob will, on a given record, always produce the same reading on the sound meter; the same machine and the same record will give the same reading if transported to Australia. Yet my 'knowledge' of the sound produced is different from my son's. He enjoys it; I cannot stand it. This then is a subjective approach to 'reality'. (It does not end there because there is the quality dimension as well: were the high decibel level to be Palestrina rather than Pop then my tolerance would be higher.)

Any study, further, can be **reductionist**[12] in its approach, as distinct from **holistic**. The first term advises us that a complex thing or process can be understood if we learn more completely about the behaviour of its constituent parts. We can fathom the nature of biological cells by investigating molecules; if there is still more explanation required then the atoms which constitute the molecules must be studied. An enquiry into sub-atomic particles is the next (and currently final) step. In this mode of thinking, for example, human personality will eventually be explicable in terms of biochemical processes. **Holism**,[13] as the word implies, accepts a whole *in toto* as the subject of description and explanation and indeed asserts that wholes have emergent qualities which are more than the sum of their parts and which cannot be predicted from a knowledge of those constituents. Holism asserts that human personality will never be predictable from biochemistry, just as the freezing properties of water are not predictable from a knowledge of the properties of the oxygen and hydrogen atoms.

Having set out these lineaments of description, we can now turn to the main constructions of the environment to which they give rise today; they will then each form a subject for more discussion in a separate chapter.

The natural sciences and technology

As now used in English, the word 'science' is the outcome of a process of progressively greater restrictions in meaning. Having started out by referring simply to the state or fact of knowing, it has come to apply to a systematic set of practices applied to a particular set of phenomena.[14] Investigations which use as much scientific methodology as possible but which look at social entities rather than non-human phenomena are qualified as 'social science'.[15] The practical applications of the natural sciences via human-constructed machines is labelled 'technology' and so embraces professions like engineering.

All the natural sciences deal in some way with the phenomena of the environment as perceived by the human senses but all have become sub-divided into specialisms. Indeed, because of this fractionation there were attempts in the 1960s and 1970s to create a new field, that of Environmental Science, which would unify all those other parts of the natural sciences which investigated the environment. If the social sciences were brought in as well, then the term Environmental Studies was often used. Although this new field proved to be a convenient unit around which to organise courses at various levels of education, it has never developed a clear methodology of its own: the sum of the parts seems to have remained just that. Another attempt at a synthesis with an environmental focus is the notion of Landscape Ecology; it is probably too early to deliver any judgement on its eventual contribution.[16] Two established disciplines, Geography and Anthropology, have long-standing interests in the relations of humans to their surroundings, and both have workers within them who participate fully in the methods and aims of the natural and the social sciences. Yet, possibly because of other concerns, neither has become the central field for scientific perspectives on the environment.

Hence the scientific discipline that has so far contributed most to the study of environment is Ecology,[17] though its position is perhaps being challenged in the 1990s by those physicists and chemists who are applying themselves to making global models of the atmosphere and oceans. The usual definition of Ecology is the science of the relations of living organisms with each other and with their non-living environment. Modern Ecology is especially concerned with the dynamics of such interactions in terms of, for example, population ecology, ecological energetics and the cycling of mineral nutrients at a variety of scales. These and other measurable parameters are held together in the construct of the ecosystem, which is a spatial area (at any scale) in which interchange of energy and matter takes place between living and non-living things. One feature of interest is the presence or absence of equilibrium (again at a variety of spatial and temporal scales) within and of an ecosystem. The study of the Earth as one single ecosystem finds one expression in the idea that the biophysical systems of the planet maximise conditions for life: the Gaia hypothesis of J. R. Lovelock.[18]

In their purest forms, the natural sciences are solely an intellectual exercise. It seems unlikely that, for example, the evolutionary history of extinct hominids or the five-toed sloth, or the counting of the number of insect species in Colombia, or the history of the population explosion in kittiwakes after 1945 are of much interest except in terms of intellectual curiosity. Yet the natural sciences are constantly interfacing with practical approaches to the world. The first of these is technology, where this is

5

defined as any humanly produced means of altering the world, be it mechanical, chemical or biological. These are of course the ways in which environmental impact as commonly understood is produced. Technology has now a special place in any discussion of the meaning to us of environment. We have used it to change so much of our surroundings that it is rarely correct to talk of the 'natural environment' any longer, and its observable and quantifiable role is not in doubt. Beyond that, though, lie interesting and difficult questions as to whether technology changes our whole outlook at fundamental levels beyond the simple 'can we build a road here?' type of question. Such enquiry becomes more a matter of discussing whether technology has acquired the power to determine ideas, beliefs and myths to such an extent that all our thinking as well as our activities are now situated within a technological context. We live, if such an argument is accepted, in a technological environment not a natural environment. At an extreme, therefore, some argue that humans delude themselves if they think they are still in control: modern technology may have acquired an autonomous character.

A second major idea proposes that science shows us how we ought to behave, i.e. that ethical principles can be derived directly from the findings of natural science; of special interest to us here are the recent history of Ecology and also the thermodynamic aspects of physics and evolution.[19] We have to see if ethical principles can be 'read off' from scientific narratives in a direct manner: a kind of linear model has often been assumed. This involves the provision of 'hard' information by the natural sciences to which the rest of society (via, for example, the studies of the social sciences) has then to adapt. Science becomes a kind of fundamentalist knowledge and like any fundamentalism needs questioning.

The 'objective' human sciences

These studies are founded upon the conviction that the methods of the natural sciences have a primary place in studies of human societies. This view in turn rests on certain assumptions such as the notion that every event has an identifiable cause and that a particular stimulus will, under given conditions, produce the same response from people. It must also be agreed that there exists an external world of actual behaviour and the results thereof which can be observed and recorded by agreed methods, and that there are indeed detached observers who can scrutinise events and processes without changing those events and processes by the act of recording them. A last assumption is that there is a structure to human society which is capable of being described in terms of regularities; this is perhaps equivalent to the assumption of order in the cosmos inherent in the natural sciences. More practically, there is the presumption that the theories developed in the social sciences can be used to change societies in predictable ways; that social engineering parallels civil engineering.

6

Hence, the aim is to elucidate regularities in human behaviour. The individual human must therefore be subsumed into studies of patterns of behaviour: the concept is called behaviourism. Classical examples of social sciences of this type have been the disciplines of Economics and Politics. In modern times their study has tended to move apart, whereas in the eighteenth century for instance they were contained in the same field, as Political Economy.[20] Economics, deriving etymologically like Ecology from the Greek *oikos*, a household, is the study of how humans put together distribution methods for any commodity which is in limited supply. In the case of the environment this means either the environment itself (where its totality is demanded as in the case of wilderness areas or outstanding scenery) or the materials that can be extracted from that environment, i.e. natural resources. Economics is said to be neutral between ends and is thus simply a study of means and so is separated from Ethics. In everyday life, a market-oriented version of economics is now the most important mediator between humans and their environments: the question, 'is it economic?' is the significant one in most projects.

Politics *sensu stricto* is the study of the organisation and conduct of government and in its objective sense has been concerned with describing how things are and how they ought to be, the particular concern of political philosophy.[21] It is not difficult to see that the possession of power to channel resources is a central feature of any distribution system and therefore as far as environmental constructions are concerned the older term Political Economy is the most useful.[22] It is also about the search for public support and power by groups who claim primacy for the environment in their hierarchy of values, *viz*. the Green parties and their equivalents.

Other social sciences with an interest in environmental constructions are Sociology and Geography. Objective sociological studies have mostly been of the type which delineate in statistical terms the attitudes of groups in society towards particular issues, such as the desirability of the development of civil nuclear power, or which measure the demand for alternative life-styles which might alter the nature of environmental impact.[23] Geographers have mostly been concerned with studies of perception and cognition, though investigations of the regional and national distribution of environmental changes which are intimately linked with social values (e.g. the designation of National Parks or commentaries on planning issues) are often found.[24]

While much of the work of the social sciences is about knowledge for its own sake, there is a strong element of wishing to contribute to public policy and action. The outstanding example of this is Physical (as distinct from Economic) or Environmental Planning. This seeks to array on the face of the Earth the various outcomes of economic processes and of environmental protection in an optimal fashion.

Many of the social sciences lead directly to ethical prescriptions and some thence to legislation. Litter, for example, has little ecological significance but is recognised as symbolic of an unacceptable social attitude towards our surroundings and so it is often forbidden to create it. The results vary by culture: contrast London and Singapore. So reading off 'good' behaviour from economics, political science and their ilk is not straightforward since the question is raised of how those disciplines constructed 'environment' in the first place.

Human experience as central to social sciences and the humanities

All that has been said so far makes it clear that the human mind is a node in the interplay of individuals, societies and their environments and this has attracted a great deal of investigation. The natural sciences, it is argued, overlook the cognitive links that join phenomena with their backgrounds and their history and so is always limited in its understanding.[25] Less restricted is a view in which the world has meaning for humans (individuals as well as groups can be considered here) and so each person constructs a world with a set of objects and relationships among which they live, rather in the manner of a bubble beyond their skin. Given the technology of communications today, this envelope may of course be quite large, and it is constantly changing in size. The usual term for this sum total of a person's involvement with the cosmos in which she or he resides is the **lifeworld**, which is a translation of Edmund Husserl's *lebenswelt*.[26] It is normal for lifeworlds to overlap and so the method is not confined by any means to non-replicable studies of individual humans and what they make of their world. Basically, however, this viewpoint discerns environments as parts of biographies in which we see people creating themselves.

In one way or another, all these constructions are analytical or explanatory. By contrast the creative arts 'just are', in a holistic sense.[27] 'In a work of art the intellect asks questions; it does not answer them.'[28] This is not to say that they have no public purpose; some avowedly have no function beyond that of conferring momentary interest or pleasure upon the observer but many others engage in dialectic with history or politics.[29] Some of the arts contribute more than others to environmental constructions but where they may point us most directly is away from the rather rigid conceptions of self and social identity which derive from the natural sciences and the more objective social sciences.[30]

Normative behaviour

Those aspects of human behaviour and of the nature of the world with which we have so far dealt all have one feature in common: they deal with things as they are, or were, or might likely be. We now move to a different category of behaviour: that concerned with how things ought to be and how people should behave: that of ethics.[31] Whereas most other constructions are concerned with description, explanation and prediction, this category is about rules, recommendations and proposals: the term 'normative' is applied. Although often related to particular cultures, in the environmental field it is always interesting to see if there are any universal maxims of behaviour that might apply.[32] In effect normative constructions depend upon the disclosure and examination of **value systems** which may not always be arrayed around human wants.[33] From these origins come environmental custom in non-literate societies[34] and law in those with writing. Most advanced societies now have large bodies of law dealing with environmental matters, though there are especial problems in dealing with any entity that cannot plead for itself in court, like a tree, or with transnational phenomena like the River Rhine or the Mediterranean.

The global scale

One difference between our world and that of the past is the scale at which human societies can affect the rest of the planet's biophysical systems. There are now global-scale changes to the planet's atmosphere as well as more regional and local changes which are found in almost every part of the globe, such as accelerated soil erosion. Equally, there can be no doubt that many more human-driven processes of all kinds are becoming global in scale. Mostly these are economic and financial, where movement is notional (as of 'money' on electronic screens[35]) but they may eventually end up as physical operations, as when the UN actually acted together in the Gulf in 1990–91 and troops together with other personnel were sent from many nations to the region in an *ad hoc* alliance. A glance at any economic atlas will also confirm that many substances are sent very large distances: oil and coal are obvious examples, but think also of *mange-tout* peas from Guatemala and Zimbabwe to Europe or daffodils by air from California to Toronto.

In our present context, this has perhaps two meanings. The first, simpler, position is that in the developed world there seems no limit to our reach for resources, whether these be materials brought to us or places to which we can travel. The world does indeed seem to be our oyster. The second, less tractable, idea is that of the global village brought about by worldwide electronic communications. On the one hand this may lead to a genuine transferral of information throughout any part of the world as we see from pictures of people in remote parts of e.g. Africa watching satellite TV.[36] All these propensities are sufficiently new for it to be difficult to make valid generalisations but they must at least be watched for their possible effects upon the ways in which we construct our models of environment. When, for instance, the first pictures of Earth from space appeared, there was a lot of comment to the effect that we would now all realise that we were indeed members of 'One World'. Has that happened? If it had, or was in train, how exactly would we know it to be the case?

Localised worlds

A movement from a world in which there were certain received truths that were more or less universally applicable to one with no such acceptances has been postulated as characteristic of the present time. The modern world that started with Copernicus (1473–1543) and Descartes (1596–1650), for example, affirmed the creative powers of mankind and has approved their expression in various universal ways: modern transport and communications, and the ubiquitous shape of the 'skyscraper' buildings of large cities are examples. In general, **modernism** as an identifiable movement believes in a controlling author (architect, painter, scientist, mogul or whatever), a logical order of setting things out (words, towns, theories) and in a metaphysical order behind it all, even if this only goes as far as the belief that the universe is indeed an orderly place. This latter school of approach bears the label **postmodernism**. It has come to have a particular set of meanings in architecture and literary theory, as a reaction against modernism's totalities. Here it rejoices in the dethroning of author and originator since the 'self' as an individual entity is denied. The world is conceptualised only in words which bear no necessary relation to the things to which they are attached. In this deconstructive postmodernism there is nothing real nor purposeful and, in terms of value, everything is relative.[37] There is, however, a wider meaning in terms of a bringing together in new relationships of the intuitions of science, ethics, aesthetics and religion, in which science and its achievements are not rejected but equally are not allowed to dominate any construction of the way the world is and how we might behave in it.[38] Chapter 6 will return to this kind of model in the light of the intervening discussions.

Order of play

Our constructions will follow a particular pattern. We begin with the natural sciences and the 'realism' which they espouse and follow them with those of the human sciences that aspire to the level and type of certainty which the natural sciences aim to provide. Then we move to the world as seen from the viewpoint of the individual human person, both in terms of the specifically environmental investigations that have been pursued and in the more implicit constructions given to us by creative artists. Then we look at what we ought to do, in both philosophical and legal terms. We end, not with Conclusions, but with an attempt to put many of these issues in a broader perspective still.

THE HISTORY OF ENVIRONMENTAL CONSTRUCTIONS

It is unreal to pretend that the constructions current today have no antecedents. 'Man does not understand himself through some kind of introspective brooding. . . . It is only by understanding the historical reality which he has produced that man becomes conscious of his power,' wrote one nineteenth century social critic.[39] Every pathway has led from a historical past of varying duration in time and space, and has been developed in a particular place or places during that evolution. So in each chapter there will have to be a certain amount of history; here I make an attempt to provide a simplified overview of the development of ideas in this field.

Not all ideas about, or developed constructions of, the environment are written down. Many pre-literate societies have had highly developed constructions of their environment which were traditionally transmitted orally. Non-western constructions may not yet have been articulated in western languages where they are accessible to European scholars in the kind of detail which is demanded. The equivalent of C. Glacken's immense compilation and analysis of the history of environmental ideas in the West before the nineteenth century has not been produced for China, for instance.[40] In operational terms, the ideas of the West are paramount today.[41] We cannot yet tell whether the world will all convert completely to the western world-view,[42] whether the ideas and constructions of other cultures will re-emerge or whether some form of amalgam (or even synthesis) will take place.

Before the twentieth century, the use of the word 'nature' as equivalent to this work's adopted employment of 'environment' was normal. Interestingly, both words have acquired many different meanings ('human nature'; 'the computer environment') but nature like environment has a dominant strand in which it means our non-human surroundings. With that connotation, 'nature' will sometimes be used here as a synonym for 'environment'.

Literate constructions in the West

This well-documented history seems to be underlain by two basic notions, out of which have grown a number of environmental constructions. The fundamental ideas are:

1 That time is linear and non-repeatable. As Eliade showed,[43] in many simple societies time was renewed each year; there was an annual cosmogony in which all things were made new, especially the ruler of the group since she or he probably had divine status.[44] A crucial change was made by Judaism and Christianity in promoting a contrasting interpretation of time as linear and non-renewable. Among other things this led to the notion of continual progress, and to the possibility that there would be enough time for the perfectibility of the world. It also allowed the flourishing of the Epicurean idea that the world was full and in decline, a view only overcome in any large measure during the Renaissance.[45]

11

2 Dualism. This emerged in Classical times and has been present ever since.[46] Simply put, the human species is different from everything else on the Earth and thus acquires certain rights and powers over everything else. Occasional forays into a metaphysical equality of all phenomena have taken place (e.g., St Francis of Assisi, 1181–1226) but never seem to have taken root.[47] Dualism was very firmly entrenched in western philosophy by René Descartes and then became an inescapable element of the work of August Comte (1798–1857) which is at the heart of today's scientific methodology. It finds its most explicit form in the phrase 'man's conquest of nature'.

These two themes have been used for a number of environmental constructions in the past. Some separate units can be distinguished:

1 Environmental determinism. This centres on the proposition that the environment is a primary determinant of the affairs of human individuals and societies. Climate is a particularly strong element in the interaction and leads to interpretations of, for example, the early rise of industrialism in England or the contrast between the economic fortunes of northern and southern Italy. It is now thought to be rather crude (and liable to misuse for political ends) but has had something of a revival in the shape of 1960s and 1970s environmentalism or Green movements, where the concepts of global environmental limits to growth of population and resource use were stressed.[48]

2 The second world. Here resides the construction of the world of mankind within the world of nature: the natural world provides, so to speak, an envelope but within that, humans rearrange the parts as they wish.[49] Though older than the medieval period in Europe, it had a powerful lift from the Benedictine movement, which saw as part of its mission the perfection of a creation which still had in it some rough places like marshes, moors and deep forests. The notion stresses creative transformation of nature and the idea of careful stewardship of resources. We may assume, further, that human institutions thus created express at least an implicit philosophy of nature.

3 The conquest of nature. When the inhabitants of this second world acquired the advanced technology available from say the seventeenth century onwards (and especially after the Industrial Revolution), and coupled that with the cultural licence to deploy it,[50] then it appeared that nature or the environment no longer posed any obstacle to human achievement.[51] Natural systems could be replaced by man-made ones which were in general superior. Today, for example, the city exhibits more order to many people than a forest wilderness. Some interpretations of Marx, as well as of nineteenth-century capitalism, propound this view.

4 As a reaction to most of these last three sets of ideas, there is the concept of a Golden Age some time in the past when all was lovely. Each period

12

of history seems to have regarded part of the past as an epoch of that kind and at present it may underlie some of the appeal of environmentalism and Green politics. In the Golden Age, harmony between humans and nature was prevalent and people sang as they worked. R. H. Grove has discussed in some detail the 'full flowering of Edenic thinking' in the mid-seventeenth century and its encounter with the reality of ecological change caused by colonial policies on oceanic islands.[52]

Clearly no one of these constructions existed in isolation for very long, and combinations of them are possible, e.g. the idea of a Golden Age will sit alongside any of the others, perhaps growing in force as the human-made extends at the expense of the natural.

Literate constructions in the non-western world

Before the dominance of western ideas and political empires over much of the globe, literate societies in several countries were often eloquent on the subject of the non-human world, whether as an object of worship, of pleasure or of utility. India, China and Japan are the outstanding examples, though only for the last-mentioned do we have detailed interpretations available in western languages.[53]

In the pre-industrial culture of Japan, a non-dualistic framework prevailed. One strand of thought in it encouraged people to value the quality of their surroundings to a degree probably unsurpassed by any other human group; the Japanese word *fudo* has its nearest European equivalent in 'natural environment' but in fact it refers to a situation in which nature and society are inseparable: 'la nature est pour les Japonais, par conséquent, peut mener à connaître ce qu'ils sont eux-mêmes', says A. Berque.[54] Whether such a world has disappeared for ever under the impact of industrialisation or whether it is only temporarily covered by a thin veneer of western ideas remains to be seen, but the existence until the mid-nineteenth century of a set of constructions different from those of the West cannot be doubted. That something similar prevailed in China is also true,[55] but in spite of herculean labours by Sir Joseph Needham and his co-workers,[56] no specifically environmental interpretation of its civilisation has come forth.

In India some attempt has been made to start to codify the environmental implications in Hindu culture and especially in the Vedic literature.[57] In some ways the attitudes so far isolated for discussion resemble those of some forms of Buddhism, which is not surprising, given the origin of this latter faith. The image of nature, for example, as an interconnected web is common to both Buddhist and Hindu.

Such over-simplified fragments are enough to give us the beginning of a sense of perspective: that though the implantations from the West may now fill the flower-beds, there have been other plants in the garden. We should encourage their continued cultivation lest our own blooms wither too early in the season.

Non-literate customs and traditions

An environmental construction, often of a normative kind, is frequently embedded in orally transmitted tradition and custom. One of the best documented is that of the aboriginal human populations of North America. A very wide gap existed between the constructions of the Indians and those of the European colonists. To the latter, this continent was often a threatening and untrodden tract empty of human life other than that of the 'savages', or possibly a sublime solitude waiting for the axe and the plough. Luther Standing Bear articulated it differently:

> Only to the white man was nature a 'wilderness' and only to him was the land 'infested' with 'wild' animals and 'savage' people. To us it was tame. Earth was bountiful and we were surrounded with the blessings of the Great Mystery.[58]

The actuality of the environmental tenderness of the Indian has been a subject of scholarly controversy since it may share with some of the written traditions a dissonance between what people wrote or said and what they actually did, a theme explored for China by Yi-Fu Tuan.[59] More precise explorations of this tension need writing for other areas of the world, filling in detail the outlines of pre-European resource use practices sketched for us by G. A. Klee.[60]

The relevance of myth

The word 'myth' is used now with more than one meaning. At its commonest it is a synonym for 'legend' or 'an old story', as in the 'myth of Arsenal's invincibility'.[61] But a more precise and historically validated meaning contains the notion of an age-old and strongly held belief, expressed poetically. J. R. Short puts it comprehensively:

> Myths are (re)-presentations of reality which resonate across space and over time . . . which are broad enough to encompass diverse experiences yet deep enough to anchor these experiences in a continuous medium of meaning. . . . An environmental myth can contain both fact and fancy.[62]

Thus the accounts of the Garden of Eden in *Genesis*, for example, are often held to be versions of the origins of agriculture and pastoralism in the ancient Near East. On entering North America, the European settlers needed to construct a myth of the land as wilderness in order to convince themselves that the natives were either not there or of no great account.

14

Myths about the environment may be one of the inputs to our construction of it, whether at an individual level or in the shape of a national (or indeed transnational) ideology. The pervasive idea of the frontier rolling forward into the wilderness is still alive in North America, for example, in the land use policies for 'wilderness areas' or in the designation of outer space as the 'high frontier' during the Kennedy administration. In Britain, 'the countryside' is still regarded nationally as a repository of certain types of scenery and environment as well as the place where particular types of people live.[63] No matter that the landscape has been changed out of all recognition by intensive farming and is lived in by middle-class commuters or second-home owners. In Australia the outback combines some of the features of both North American and British themes. One of the enduring myths of the city in the West has, of course, been the place whose streets were paved with golden opportunity, as well as the earlier ambiguity of being at once identified with heaven-on-earth ('till we have built Jerusalem . . .') and as a place of danger and iniquity.

The ways in which environmental myths can be constructed are as various as the channels of communication in any society. Oral transmission round the fire at the end of the day may be paramount. If the fireside is flanked by a TV set which is normally switched on, then the environmental myths come through programmes, news bulletins, advertisements and all. That a myth may at any one time not be consonant with what the natural sciences tell us does not mean that it has no power: if science is one truth (in that context), then myth turns out to be somebody else's truth. If we adopt a myth that the environment is an inexhaustible cornucopia of resources provided that science and technology can develop untrammelled, then that has for some as much authenticity as any evidence that such use of the environment has destabilising effects upon the biogeochemical systems.

Myths can be one class of interactions between human societies and their non-human surroundings. Luhmann points out that ecological systems cannot communicate directly with humans and so our cognition of these systems and their behaviour is a result both of such direct contact as we may have and of all the other channels of communications about environmental matters.[64] These no doubt include those which are historically derived, like myth, together with those which we do not think about consciously since they are so deeply embedded in our world-view. These cognitions set up what Luhmann refers to as **resonance**. The changes in the real world, together with our cognitions of them set in motion a set of vibrations through society's channels for discussion which cause anxieties and result in laws, environmental education and possible changes in world-view. Hence how society structures its capacity for processing environmental information is critical. One fundamental aspect of the structuring is that it is formulated in a series of binary codes. These codes (legal/illegal; possession/nonpossession; conservative/progressive; restrictive/expansive) claim universal validity and exclude further possibilities. Whether this is a healthy way of approaching our construction of world-views and environmental meanings is discussed further in the last chapter.

THE EMERGENCE OF PROBLEMS

We would not now be examining closely the *ideas* surrounding environmental change if we were not concerned in some way about it. That concern focuses, of course, on another construction which is probably common to all human societies since time became linear: a concern for the future beyond the immediate short term.[65] In the present context, this concern has two main strands:

1 Environmental degradation. This is a cultural valuation at local and regional levels, dealing with the levels and kinds of environmental manipulation and contamination that are acceptable. It changes gear somewhat when we are forced to consider global transformations such as the enrichment of the atmosphere by radiative forcing at the same time as measurements of the ozone layer in the atmosphere suggest that it is being attenuated. The level of concern becomes global when the results of some scientific studies of the atmosphere are evaluated. This is especially so with those which concern 'greenhouse gases' and ozone depletion since the models predict considerable consequences for most of human society.

2 Resource depletion. This is a highly complex matter since it involves not merely the physical amount of the resource but its price and its cultural acceptability, which is very important in the case of food, for instance. But the high material expectations of many societies have led to immense demands for energy and materials at rising per capita rates of use. Many writers have argued that the world cannot sustain these burgeoning rates for very long: there are differences between long-term stability of some ecosystems and the short-term demands of some human societies. We have to consider what adaptations of ourselves and our technology we will make to avoid unacceptable levels of material deprivation. These could well lead to global warfare, with all the far-reaching environmental as well as social consequences of that event. This concern links in with the first since much environmental manipulation takes place in the cause and course of garnering resources. Indeed, for some protagonists, 'environment' should be replaced by 'natural resources'.[66]

There is inevitably a diversity of attitudes towards what are defined as problems. However, it does seem unfair to talk of 'environmental problems' as if 'the environment' were itself somehow acting wrongly. Whatever it is, some or all of it was there a long time before us and our ways. If there is disharmony between humans and the rest of the Earth, then it seems more likely that there is a set of social problems which have, in part at least, an environmental delivery route. In which case we need to be searching for the knowledge desired by Jan Amos Comenius (1592–1670), 'universal, disgrac'd with no foul Casme'.[67]

2

THE NATURAL SCIENCES AND TECHNOLOGY

The material which deals with the most relevant areas of science and technology[1] for our present purposes is presented here in a series of sections:

- The general nature of the natural sciences and of technology.
- A discussion of the particular natural sciences with the most to tell us at present about the environment.
- Can we read off directly from science and technology what behaviour we should adopt towards the environment?
- Are there deeper philosophical and practical questions which we ought to carry forward to any final discussions?

PURE SCIENCE

Science has a reputation for producing knowledge of an altogether different quality from any other methodology; indeed it has been elevated by some of its practitioners to the heights of being the only knowledge worth having. There are, though, calmer views of the kind of knowledge which science gives us as well as radical attacks on both its philosophy and pragmatics.[2] 'Science' is now taken to mean 'western science'; however, until the Late Renaissance, that of China was as advanced as that of the West in, for example, astronomy, algebra and quantitative cartography. It was allied to a technology which could produce the development of water-power, the ability to drill deep holes for brine and natural gas, the foot-stirrup and the wheelbarrow.[3]

The present power of pure science has come from its representation of **consensual** knowledge: statements are tested and agreed independently of any personal characteristics of the observer. In its most ambitious form this requires a totally unambiguous language, the provision of which is one of the aims of mathematics. In essence, the scientist as an observer and as a language user can capture external events in statements or messages that are true if they correspond to the facts and false if they do not. Any true statements or propositions thus have a one-to-one relation to the facts even

18

if the facts are not directly observed. This latter condition may result from hidden events or properties or events which are distant in space or time. Such events are described in *theories*[4] which are inferred from observation: it is assumed that hidden explanatory mechanisms can be discovered from what is open to observation. In all this process, humans can stand apart: there can be knowledge without a knower or without a thinking subject.[5]

Science is above all **empirical** in the sense that it starts by making observations which are based on the experience of the senses, an experience which may be much enlarged by technological means. Science tries to uncover facts from a distanced and external world: the development of empiricism which is called **positivism** avers that all assertions not subjected to scientific norms of testing are not related to truth. Clearly, the validity of any empirically based truth must depend upon the quality of the observations and there is a lively debate about whether really objective observations can be made or whether they are all preceded by theory. Thus one viewpoint argues that the mind is not an empty computer memory in which observations can simply be stored but has of necessity a pre-existing program which sorts and orders the observations. Scientists will, however, argue that controlled observation and experiment can bypass many of the idiosyncrasies of human cognition. These procedures are, however, practical achievements and failure is possible.[6]

Science does not necessarily subscribe to the idea of a universal standard of rationality with respect to which one theory can be judged better than another. To that extent it is **relativist**. The philosopher Ludwig Wittgenstein (1889–1951) suggested that there were only language games and that truths were relative to each of these games; there was no judgement possible between one and the other. In real-world terms, though, science proceeds as if it were arriving at the truth, making interim statements along the way, even though there is no necessary convergence of the conceptual framework of theories towards a universal truth. Equally, we cannot deny that it is progressive in a practical way (the word 'instrumental' is often used), again usually in alliance with technology.[7] One of the great powers of science has been that of prediction in time. Certain theories predict future events with what seems like 100 per cent probability: the theory of gravity is one example. In terms of the everyday world and of practical affairs, much of the alliance of science with technology rests on predictability. Sociologically too, the pecking order of the sciences in academia rests on this feature: physics usually occupies a high place for example, ecology a much lower one, and the social sciences (where the complexities of human behaviour have to be taken into account) are well down the ladder of esteem as formulated by those at the top.

Any scheme of knowledge which makes strong claims for itself is bound to come under intense scrutiny; in the case of science this is doubly so because to pursue it generally costs money for which there are competing uses. Those who claim to have found flaws in science have generally been from two groups:

1 those who claim on grounds of logic and philosophy that not all the claims described above can be upheld; and
2 those who regard science as one cultural activity among others and that when it is placed in that framework it exhibits certain qualities that societies ought to watch very closely since dangers might ensue from the continuation of such activities.

An example of the first category is the assertion that every statement of observation has to be expressed in a language which rests in turn upon certain other theories. Thus they can be only as precise as the founding theory is precise: 'force' in physics is precise; in meteorology ('gale force winds') less so, and in rhetoric ('the force of her argument') even vaguer. If all observation statements presuppose theory, this would be important for environmental observations.[8] The second category is simpler and will be mentioned at appropriate points. It goes rather beyond the simple excoriation of chemists for inventing CFCs to the notion that the whole of science is an instrument for domination: of one group of humans by another or of the environment by humans. In such cases, prediction would be an essential tool.

Technology

Technology is not easily differentiated from science. It has often come first in the sense that machines have been made to work without the inventor knowing much about the precise value of the gravitational constant or the laws of thermodynamics. On some occasions technology clearly addresses a different audience from science: problem-solving in the short term is more important than elegant mathematics and there is more advocacy allied to less scepticism. Its successes in providing material benefits for those with access to it are undeniable. Technology will work only if conditions A, B and C are satisfied. Therefore it is attempted only if conditions A, B and C are present. If they are not, then it may become essential to change X, Y and Z into A, B and C. In this view of technology (with many implications, of course, for environmental change), technology is a neutral tool to be used for human ends. These may be good ends, as in boring wells in dry lands, or bad ends, as in slaughtering migratory birds when they cross Italy.

This view of technology as a kind of impartial instrument is not universally held. Some think that there is very little control exercised: the test for society is, simply, if you *can* make it then you will, as with Concorde and nuclear weapons.[9] Those who innovate are remembered: it is the stoppers who are forgotten. These examples lead to the opinion that some technologies are inherently political: some lead to centralised authority, others to decentralised decision-making. In 1872, for instance, Engels repeated the example put forward by Plato that a ship at sea must have a single captain and an obedient crew; both writers extrapolated somewhat from that example. But the view persists: if you have a given technical system, runs the argument, then you have to have a particular set of social conditions: it is impossible to run a railway system or an oil company as a set of small family businesses.[10] If you have a lot of plutonium circulating in an economy then government infringement of rights of privacy must be allowed so that any theft of plutonium can be rigorously followed up. This leads to the belief that technology begins to shape human affairs, like where the understanding of technology has merged with self-understanding, as in our use of terms like 'interface', 'network', 'output' and 'feedback' and especially in the common comparison of the human mind with a computer. It steers us towards the possibility that technology can become a master rather than a servant.[11]

The closest science to matters of the natural environment seems to be biology, no doubt since living organisms act as integrators for a large number of other factors from their surroundings, as well as of history. The branches of biological science which deal with (a) the relationships of living organisms to each other and to their environment, and (b) with the history of the existence of species are, respectively, **ecology** and **evolutionary theory** and it is to these and to one synthesis of them put forward by J. Lovelock and known as the Gaia hypothesis that we shall now turn. On the way we must consider the problems raised by the new generation of modellers of climate who are interested in prediction on all manner of temporal and spatial scales.

ECOLOGY AND CLIMATOLOGY

If science has indeed regarded as fundamental the evidence of the human senses, then it is not surprising that the nature of living things and the weather have been leading elements in scientific constructions of environment, though of course they are not the only ones. For many years, both biology and the atmospheric sciences were concerned with data accumulation and classification but in their modern phases they are both involved extensively with the dynamics of process. They both recognise, to put it very simply, that explanations derived from the study of static patterns (like plant distribution or continental rainfall maps) have severe limitations and that closer investigation of what happens between these time- or space-slices is essential.

Ecology as a science

Although there is occasionally some reticence on the part of ecologists to commit themselves to a definition of their discipline, we can perhaps be content with the commonly accepted notion that it is:

the scientific study of the relations between plants and animals and their non-living environment.

We should note that humans *per se* are not mentioned and we shall have to return to this matter. More important at the moment is the idea that the definition is scale-free as far as space is concerned. Ecology can set itself to investigate the heat exchange properties of a small beetle living under the bark of a dead tree just as well as evaluating the role of living organisms in the total flow of carbon around the planet. Both can, of course, be difficult: nature does not divide itself into neat hierarchies of spatial regions like government departments. There are invariably problems of complexity since the scientist's normal reaction is to look for explanations at the next lower level of organisation (which may in this case be a smaller area or a smaller set of organisms), in the classic reductionist mode. Yet it is not at all certain that events at one scale translate simply into those at a higher or indeed lower level of organisation and complexity. So knowing at what scale to observe an ecological entity can present difficulties.[12]

Time is not neatly defined either, though it is customary to separate ecological time from evolutionary time,[13] with the last 10,000 years being given over to the ecologists. This gives plenty of years for processes to occur, and significant events may occur at possibly 50- or 100-year intervals. The explosion of the crown-of-thorns starfish in the Pacific was attributed to many causes before it was realised by coring coral reefs that it was a periodic event of some antiquity; the 1970s outbreak was the first since scuba-diving became popular.[14] Given the complexities inherent in these rather ragged edges of time and space, it is scarcely odd that ecology should not have acquired a good reputation for making accurate predictions. J. Maynard Smith, an evolutionary biologist, is quoted as saying that ecology is a branch of science in which it is usually better to rely on the judgement of an experienced practitioner than on the predictions of a theorist.[15] But not all areas of physics can do more than talk of the probability of alternative outcomes, yet in the case of the ecology of fisheries, for example, the effects of different types of exploitation can be foretold with a considerable degree of biological and economic precision.

Although the coining of the word 'ecology' is usually attributed to the German biologist Ernst Haeckel in 1866, the concepts behind it have a long history.[16] In the late nineteenth and early to mid-twentieth centuries, ecology inherited some of the mantle of natural history and underwent something of a static phase when cataloguing seemed its main purpose. It was rescued from this by developments such as the mathematics of population ecology (like the relations between cyclic populations of predators and prey); the emphasis on competition and diversity which it received from evolutionary theory; and the functional approach based on the ecosystem concept, trophic levels and the flow of energy and matter.[17] These fed into a more

22

general model of a self-regulating set of changes moving in the general direction of greater complexity of organisation which has been fundamental in studies of ecosystem stress and stability.[18] But more recent work has spun apart this rather neat unidirectional sequence and its implications for environmental management. Empirical work on successional change in forests in the USA suggested that there is no determinable direction of change and that it goes on without reaching a point of stability. There were none of the expected features predicted by some ecosystem theories such as an increase in biomass or diversification of species. A forest is a shifting mosaic of trees and other species without an emergent, holistic collectivity except in the physiognomic sense.[19] The result, for our larger theme, is acceptance of the idea that disturbance and readjustment are continual and that outcomes of change are very poorly predictable.[20] In this respect, these ecological views look forward to the chaos theory.[21]

The ecology of humankind

In this section we shall deal with ecological interpretations of the position of mankind in ecological systems, the notions of ecosystem stress and stability, and ecologists' approaches to ecological problems and solutions. Their approaches to the role of humans in ecological systems have changed in recent years. At earlier stages, writes W. A. Reiners:

> it has been customary to view the human-dominated world as a harsh and strange place for the native biota because of the predominance of disturbance associated with human activities.[22]

In this view, the nature of human consciousness was a differentiating factor which separated *Homo sapiens* from all other organisms and so people became an 'outside' or allogenic influence upon ecosystems. Later thinking then suggested that even unpopulated landscapes were subject to disturbance and ecological stress: that more of a continuum existed. The ecosystem approach of H. T. Odum and E. P. Odum,[23] with its emphasis on flows of energy and matter, has allowed the incorporation of human activity since measurements of the deflection of energy flows, for example, or of the import of energy from fossil fuel sources, can be plugged into the type of systems formulation adopted (Figure 2.1). Human societies thus become 'part of' rather than 'outside' the systems under study. In biological terms, *Homo sapiens* can be seen as, for instance, an outside influence, rather like a sudden change of climate or a natural disaster; or, humans can be characterised as a component of the system, indeed often as the dominant organism whose influence determines the nature of the rest of the system as do the trees in a forest.[24] If a more humanistic perspective is required, then features such as information transmission can be added to energy and matter flows as elements of the system's functions.[25] Finally, a whole human

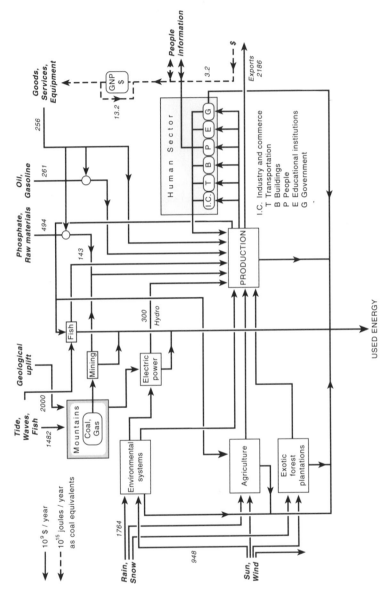

Figure 2.1 An attempt to model the ecology of New Zealand by bringing into one set of flows both energy and money. Though the equivalences are subject to debate, such diagrams encourage integrated thinking about what 'an economy' really means

Source: H. T. and E. C. Odum, *Energy Basis for Man and Nature*, New York: McGraw Hill, 1976, Fig. 10.7, p. 161.

culture can somehow be worked into the other linkages (e.g. as culturally determined choice-points) to try to encapsulate the variety of aims which humans may have in bringing about alterations in their surroundings.[26]

Not surprisingly, the term 'human ecology' flickers in the background of many of these discussions. In much of the twentieth century, biologists cut themselves off from the species Man as an object of their studies, but more recently attempts to reclaim this territory in terms of evolution and behaviour and of history have been made.[27] Another field where the effects of humans and those of nature run closely together is that of stress ecology.[28] Such studies aim to measure and evaluate the impact of natural or other external perturbations both on the structure and function of ecological systems and on the management of such systems, either for the perpetuation of the species themselves or for the benefit of humans. Inevitably, prediction is involved, and this always becomes more difficult as the gradients of time, space and biological complexity get larger (Figure 2.2).[29]

Ecology and problem-solving

Environmental 'problems' are not new. They have appeared in the literature of the West and China since writing began, and doubtless any fully recorded

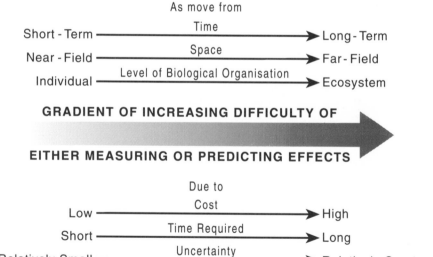

Figure 2.2 Gradients of time, space and biological complexity which contribute to the difficulty of measuring impacts upon ecosystems and to the predictability of such impacts

Source: G. W. Barrett and R. Rosenberg (eds), *Stress Effects on Natural Ecosystems*, Chichester: Wiley, 1981, Fig. 3.1, p. 32.

oral tradition would contain many examples of the disappearance of animals or the loss of pasture or of soil. The great landmark volume appeared in the nineteenth century: G. P. Marsh's *Man and Nature*,[30] which, although anticipating the word 'ecology', has such a strong sense of the interconnectedness of phenomena that it is usually claimed as a pioneer of the ecological approach to environmental problems.

In the 1960s and 1970s, the multivariate nature of processes such as the scarcity or imagined future scarcity of natural resources and the perceived degradation of environment caused by contamination with various toxins, led to a great surge of interest in ecology. It was seen as a science which might explain the reasons for such difficulties and, further, lead to policies which would remove the anxieties that underlaid what was then generally labelled as 'the environmental crisis'. Although ecologists had often warned about the deleterious consequences of certain human actions, some were very cautious about their new status as prophets *sensu Hebraica*. To be sure, one or two embraced the role of jet-set Cassandras and others achieved the high standing of those called upon to sit on blue-ribbon committees in national capitals or to issue rentaquote statements to the media every time a duck fell dead. Many professional ecologists, however, felt that since ecology dealt to a great extent with uncertainty that they had to make the probabilistic nature of their predictions a feature of any involvement with public life. They pointed out[31] that in any ecosystem there was natural variability, of which cyclic fluctuations of animal populations were the simplest example. Nevertheless, ecologists became involved in policy-making. They saw that much information already known was not finding its way to legislators and that quite possibly the perceived integrity of scientific data could be preserved during its passage from the field, through the computer to the specialist report and thence to the committee hearings, the TV appearances, the making of law and its day-to-day enforcement. 'Scientific data must be evaluated, authenticated, organized, analyzed, and interpreted for use by legislators and administrators', wrote one North American[32] in 1970.

The longest-lasting outcome of these trends was the enactment in the USA in 1969 of the National Environmental Protection Act (NEPA), which called for Environmental Impact Statements (EIS) for federal projects above a certain size; this Act has been imitated now at other government levels and in many other countries.[33] Its basis was that of carrying out an Environmental Impact Analysis (EIA) in advance of any major project, and the terms of reference of the EIA were usually ecological in the widest sense, i.e. ecosystemic rather biota-oriented. (Some versions go further and call for Social Impact Analysis as well.) A great number of these EIAs have been carried out in many parts of the world: most, alas, remain unpublished in the open literature. In spite of their widespread use, a number of criticisms have been put forward by the scientific community. One of the problems is often

26

the language used in legislation: this may well not translate easily into scientific equivalents. One set of regulatory standards in the USA, for example, asks whether a species is 'important', which may be definable in commercial or recreational terms, and may readily be demarcated as rare or endangered. But it is another domain altogether to say whether or not 'it is critical to the structure and function of the ecological system'. Refining the techniques involved is no doubt possible but the major obstacle to science, that of persuading those who have political power to believe in the results and to act on them, will no doubt remain intractable. But the progress towards the control of nuclear arms signalled by the signing of the INF Treaty in 1987 can be viewed, at least by some, as a response to the global EIS-type scenarios of the 'nuclear winter'.[34] The differential response of national governments to predictions of climatic change from radiative forcing is another example of a non-linear relationship between scientific data and public policy.

Climatology

In 1960, there were probably less than 50 climate modellers in the world. Now there are several thousand of them and they have a dominant voice in the constructions of environment being made by scientists who are called upon to comment on the likelihood of human-induced global warming. Their remit is wide: the Global Atmospheric Research Programme deals not only with the atmosphere but also the dynamics of the hydrosphere, the cryosphere, the land surface and biomass. The aim at present is to develop and refine models of the atmosphere of different types and to end up with powerful and predictive Global Climatic Models (GCMs). One type of model is basically numerical and the other is analogue. The latter uses historical data to try and assess conditions on the earth's surface when, for example, carbon dioxide levels were different from the present, and to do the same for feedback effects linking cloudiness and temperature.[35] Another important input is the amount of plant material on the globe. This is critical in absorbing carbon dioxide, and is measured from satellites in the form of a Global Vegetation Index or GVI.

As in ecology, there are problems of dealing with the very large data sets that can be acquired (especially from satellites) and so there are problems of archiving information and gaining fast access to it. Equally important, there are intellectual problems with the data and their use: even with all that is collected, GCMs are still plotted on a coarse spatial basis, for example. And so far as prediction is concerned, the forecasts are never any better than the assumptions that are made about the relationships between past, present and future.[36] Even short-range weather forecasting, we may remember, has quite severe limitations. So both ecology and climatology deal, in their modern forms, with dynamic systems of great complexity. Ecology was elevated to be the number one science during the 'environmental crisis' of 1965–72, and climatology now occupies that same position. In both, it is the interaction of humans with the biophysical systems of the planet that is seen as a source of anxiety. In both, we might add, humans are often seen as disturbing a former state of equilibrium which is usually evaluated as benign. Ecology has moved from a focus largely on problems at a regional and local scale to the global state of e.g. oceanic phytoplankton and the effect of raised levels of ozone in the upper atmosphere upon plant life, via the medium of enhanced levels of UV-B radiation. Climatology has been much more planet-wide in its initial scope, although the regional effects of possible climatic change and sea-level rises are also studied. In common, also, are the suggestions for technical fixes: the suggestions of sowing the seas with iron filings to encourage phytoplankton to raise their NPP levels and thus sequester more carbon dioxide, or planting fast-growing and long-lived trees everywhere to achieve the same, are such measures. Not totally serious perhaps is the suggestion for Europe that Box trees (*Buxus* spp.) should be planted everywhere since the wood is denser than water and so could be sunk to the bottom of the seas along with its freight of carbon.

EVOLUTION AND ENTROPY

This section is devoted especially to two concepts which give chronological depth to scientific constructions of the environment. The first is largely biological, namely that of evolutionary theory. The idea of organic evolution (adumbrated before Charles Darwin (1809–82) and indeed more or less simultaneously put together by Alfred Russel Wallace (1823–1913) but associated with Darwin because of his depth of evidence and cogent argument[37]) asserts that species have come and gone during the millions of years of life on the Earth. It further suggests that life forms have taken on increasing complexity during those epochs and that humans are a product[38] of the same evolutionary processes which produced, for example, the other Primates and even the AIDS virus. The idea of entropy is not so easy to comprehend. It was firstly a measure of thermodynamics and related to the Second Law, in which it becomes a measure of the disorder among the atoms whose state constitutes the energetics of the system.[39] An initially ordered state is bound to become random as time proceeds: high quality energy able to do work must ineluctably become low-grade heat unable to do so. In a broader frame, the concept has been applied to the Earth as an open system which takes in ordered energy from the sun but converts it to low-grade heat which is radiated back to space. In between, the energy has powered life and many other complex processes but in the long run, disorder and chaos must prevail: the universe will become randomised.

We have to ask whether these concepts, evolution and entropy, in any way form environmental constructions by themselves or whether they add to others; for example, do they amplify the ecological construction of environment or do they go beyond it?

Darwinian evolution

Writing of the history of ecology[40] admittedly, but taking a very wide view of its antecedents, the North American historian D. Worster asserted of Charles Darwin that 'no other individual has had so much influence generally on western man's perception of nature'. Thus we must consider the manner in which his ideas have contributed to world-views which are still prevalent today and equally ways in which his theories have been elaborated into constructions of environmental significance. Two of Darwin's books are especially important for us here: *On The Origin of Species* (1859) and *The Descent of Man* (1871), though the *Beagle* journal of 1839 is also full of insights which clearly influenced his more mature considerations.[41]

The absolute kernel of Darwin's contributions is the theory, not yet falsified by scientific evidence, that species die out and are replaced by others. As S. J. Gould puts it:

> Life is a copiously branching bush, continually pruned by the grim reaper of extinction, not a ladder of perpetual progress.[42]

The process of replacement also at one stage produced *Homo* just as it had other taxonomic groups. Out of this seed, and out of examination of the processes which result in these changes, a mighty set of constructions has grown, all the way from the purely biological to the intensely moral, and most are of interest for the present theme. For Darwin himself, the world had been a fallen place ever since his visit to the Galapagos Islands, whose bleakness impressed him greatly. One result was his view that there was no necessary law of development and no inherent tendency to perfection in nature. Once this view of the world was combined with the acceptance of man as a product of evolution rather than special creation, then it became clear both to him and especially to others that the building of 'civilisation' had to be accomplished via the intellectual and moral faculties of humans; it was not something that could be allowed to take care of itself.[43]

One result was a startling justification for a Victorian ethic of domination over nature, a situation which was becoming technologically feasible at an ever greater rate as the power of steam and iron was realised. The lesson of history became elaborated into a long ascent from chaos to managerial control of a nature which was haphazard and inefficient (witness all those spare fish eggs and the lazy meandering of rivers) and from which western men had to be distanced by a wholehearted declaration of independence. Since indeed nature was in a fallen state, then civilisation was a necessary moral check upon it. The ideas of Thomas Henry Huxley best embodied this position: for him, conscience made for a positive ethic of separation from nature so as to strive to build something better. In 1893, he delivered the pronouncement that: 'Social progress . . . means a checking of the cosmic process at every step'.[44] Among other things, this kind of view gave rise to the development of Anthropology in the 1850s, established as a means of clarifying the relationships of civilisation and savagery. Even stronger were the purely social translations of Darwinism (not dealt with here) and together they drowned out another thread of Darwin's writing, that of a more biocentric approach to nature. His words when describing some of the sights encountered on the *Beagle* voyage were far from those of conquest and his memorable phrase (in a notebook of 1837) in this context was that eventually man and nature were 'all netted together'.

The model of evolution that dominates our own time is perhaps that of a ladder or, better, an escalator in which genetic change carries *Homo sapiens* inexorably to the top.[45] The escalator model has provoked some counter-reaction, not least from those who see that rapid extinction is just as likely an outcome as some form of perfection at 'the top'. With humans, of course, the cultural transmission of acquired information becomes ever more important and there is now some tension between those who assert the primacy of the cultural and those who follow very closely in Darwin's footsteps by asserting the predominance of the very stuff of evolution in the shape of genetics. For some biologists, organic evolution itself will carry on in humans via genetic changes which 'improve the nature of man himself';[46] for others this is made self-evident in terms of extrapolation from the diminishing intervals between earlier stages of evolution; in the case of the field of sociobiology, the genes are said to hold culture on a leash, and so a genetically-controlled evolutionary function is seen as a competitor with, and indeed a loser to, psychological processes such as motivation: there is no free will here. Some commentators still hold to nineteenth-century notions that the struggle for existence affects ideas and institutions as much as it does species and so presumably cultural constructions of environment are to be winnowed out in much the same way.[47] This last bundle of ideas has many challengers.[48] Those who prefer to put their money on the features of human consciousness such as free will, and those who do not see lemons and lilies, peacocks and pythons simply as failed humans, argue for a different kind of outlook. Their environmental construction contributes to the environmentalist or 'Green' position in most debates.

We can perhaps see evolution in a broader frame along with Erich Jantsch when he talks of evolution and the future for mankind in two ways.[49] The first is evolution by extension of the environment into space, the discovery of many New Worlds. The second is the opening up of new niches for mankind by the extension of our consciousness; this can naturally include information technology though not as part of a machine-dominated drive to conquest but as an extension of personal experience. Learning thus becomes a creative game played with reality, a co-operation between autonomous wholenesses, and is infected with the idea that evolution is never purely functional: there is always some extravagance. Do we really want to live, asks Jantsch, in a predictable environment which can be rationally controlled?

Entropy and the validation of time

In our current field of endeavour, entropy is of most interest in so far as it is a measure of energy in the universe and hence on Earth. The energy comes in two states, namely available or 'free' energy and unavailable or 'bound' energy. These are often anthropocentric terms, meaning available or not available to mankind, and entropy is a measure of the conversion of one to the other. In coal there is free energy and low entropy which is converted by combustion to bound energy in the form of heat, smoke and ash, and high entropy is created. Also, the presence of free energy implies the presence of some kind of ordered structure whereas bound energy (and high entropy) indicates the dissipation of the energy into disorder. On Earth, there is free energy in the form of the rather dilute sunlight. It also comes in the more concentrated form of fossil fuels but this non-renewable stock in fact constitutes no more than a few days' supply of free energy as measured by the input of solar radiation.[50]

This teaches us a few long-term lessons with implications for environmental matters. For all isolated systems the future is of increasing entropy and this confers a uniqueness on each instant of time since entropy is continually increasing and cannot be reversed. This slightly gloomy finding is offset by the observation that in non-equilibrium states (which is clearly the condition of our planet) the flow of matter and energy although creating entropy is also a source of intricacy, i.e. irreversibility is a condition of the creation of complexity. This cosmological fact of a world far from equilibrium implies a different sort of order from that of Newtonian mechanics; throw in the indeterminacy principle of quantum dynamics and time becomes a form of construction because of the random bifurcations which must occur.[51] Closer to everyday concerns, economic processes must be seen as entropy-creating since they eventually degrade energy and matter into waste, and energy into low-grade heat, and matter into junk and refuse. No technology can at present produce its own fuel but must rely on some form of stock; no solar collector has yet garnered enough energy to reproduce itself.[52]

THE GAIA HYPOTHESIS

The term 'Gaia' can be used as shorthand for the Gaia hypothesis, which is an holistic model of the outcome of global evolution and ecology considered together and associated with the name of J. E. Lovelock. It stands conventional ideas about the evolution of life on their heads since its core is the statement that the Earth is a self-regulating entity with the capacity to keep the planet populated with living organisms which control their chemical and physical environment. Thus life did not evolve to fit the conditions of a cooling planet but adapted them to ensure its perpetuation.[53] The analogy is perhaps with the fur of a cat or a wasp's nest: neither are themselves living but they are essential for the continuation of that form of life. Gaia should not, however, be thought of as some form of sentient being; the term 'she' can be used as it would be employed of a ship, with the connotations about such a vessel being rather more than the sum of its parts.

The basic concept

The Gaia hypothesis revolves around the ideas of cybernetics, the study of control organisations. It suggests that life itself controls the systems which seek to optimise the physical and chemical conditions for its existence, though not necessarily for human life. Life produces a non-equilibrium set of states of, for example, the composition of the atmosphere and the salinity of the seas, both of which are different from the predicted levels should life be absent. It thus regulates the amount of carbon in the atmosphere by sequestering excess amounts in plants and at the bottom of the oceans and also regulates salinity by precipitating minerals in tropical shallow-water areas and to the deep ocean floors. The hypothesis is also retrodictive: in historical terms it suggests that Gaia should be able to withstand the loss of some of its parts and this indeed happened during the Pleistocene when most of the area polewards of 45° was inert under the weight of ice-sheets. Yet full biomes regained the area quite quickly after the great meltings. The predicted connection between life and the atmosphere was thought to have been fulfilled by the discovery that phytoplankton emit a gas in aerosol form which then forms the cloud condensation nuclei over the oceans. Thus precipitation amounts and locations would be governed by the activities of living organisms, but the story is probably more complicated than that.[54]

Gaia and environmental impact

During the twenty or so years of its existence, many commentators have looked for environmental lessons in the Gaia hypothesis. It seems as if the most sensitive parts of the cybernetic web are particularly the microflora of the oceans and the soils, which together with some forests are responsible for the turnover of about half the world's carbon. It is therefore human impact upon these systems which is likely to have the strongest effect upon the function of the Gaian systems: the poisoning of the continental shelves of the tropics and the felling of tropical moist forests seem to be the obvious examples of such processes. Lovelock is sanguine about industrial pollution in the temperate zones, thinking it to be analogous to the loss of living area during the Pleistocene and in any case under rich peoples' noses and therefore likely to be dealt with. In 1979, he thought that the ozone layer problem was being compensated by the carbon dioxide excesses but it now seems that these may be different in their reactions with UV light. But there seems little doubt that human impact on Gaian systems could, depending on its intensity and its location, throw the whole set of webs into a new phase of non-equilibrium, i.e. Gaia could 'flip' from one state to another rather than undergo any form of gradual transition.[55] The adaptation towards the perpetuation of life on Earth does not automatically mean human life and indeed if human activities violate the basic mechanisms then they are likely in the long run to be swept aside. The hypothesis suggests that the planet is not of itself an anthropocentric place: it could be the microbes and the amoebas that next inherit the earth.[56]

The status of the Gaia hypothesis is still open for discussion.[57] If we think of a 'strong Gaia' theme, then the Earth is some kind of superorganism: an emergent whole with qualities that transcend those of individual components. A 'weak Gaia' hypothesis would restrict the idea to a series of complex feedback mechanisms on a global scale, a kind of geophysiology. The 'strong' view has the additional and intriguing characteristic that it is not strictly Darwinian: the behaviour of the whole is more purposive than Darwinian selection normally allows.[58] That is, evolution is goal-directed: the optimal end-state might be more like a process such as the maintenance of an overall structure in the face of change or an increase of order and information. At present the lowest common denominator is perhaps a powerful metaphor, and as such by no means to be despised for metaphor is a central descriptive device in most sciences. The idea, too, of an emergent domain which links all living things and their surrounding environment, in a self-regulating system with an open but non-purposive nature, does not violate scientific norms of validity.

PROBABILITY THEORY, ENVIRONMENTAL CHANGE AND CHAOS

In this short section attention will be paid to the application of a branch of statistics to dealing in a practical way with a feature of our knowledge of the environment, namely its uncertainty. As the discussion of ecology has shown, that science has not managed to achieve anything like the success of physics in making temporal predictions, largely because it deals with indeterminate systems, i.e. those where a number of outcomes are possible no matter how similar a set of starting-points. The mathematics of chance are founded on probability theory, and strenuous efforts have been made in both science and business to apply this in their dealings with factors of the environment.

Risk analysis

There can be objective measurements of risk in human life, such as the relative risks of different methods of travel, for example. In more detail, three main classes can be distinguished:

1 risks for which statistics are available;
2 risks for which there may be some satisfied evidence but where the mechanism linking the suspected cause and the injury cannot be traced, as with latent cancers after exposure to radiation, for instance; and
3 risks of events which have not yet happened or which have come about in such small numbers that no reliable probabilities can be established.

In the first category, for instance, deaths per 1,000 km travelled in the UK by rail in 1972–6 were 0.45, by scheduled air transport 1.4, by car as driver 8.0 and on a motorcycle as a passenger, 359. Similarly, occupational risks can be calculated accurately for a whole population: the annual average accidental death rate at work in the UK per million at risk (1974–8) was 5.0 in the manufacture of clothing, 110 in agriculture, 210 for coal miners and 2,800 for deep sea fishermen.[59]

But this is by no means the same as perceived risk, which involves the latter two categories described above. Perceived risk is the combined evaluation made by an individual of the likelihood of the future occurrence of some adverse event, and its significance to the person concerned. The perceptions thus constructed may bear a rather uncertain relationship with the objective data of the type quoted above. People may overestimate the risk of rare adversities and underestimate the common ones, for example, and their idea of what constitutes an acceptable risk (and which they may wish to enshrine in public policy via legislation) may not coincide with the objective data presented by the statisticians. Concentrated and obvious risks such as major air crashes or large-scale industrial explosions are regarded as worse than diffuse risks like the normal run of road accidents. Risks to customers are less acceptable than those to workers in an industry; involuntary risks (e.g. food contamination from pesticides or salmonella) are seen as worse than voluntary risks such as riding a motorcycle or rock-climbing. Immediate risks are perceived as more dangerous than deferred hazards, and new risks achieve a much higher profile than older established ones, no matter what the actual level. Risks evaluated by groups suspected of partiality are given less credence than those regarded as having been evaluated by impartial groups. And in general, the cost of saving a named life are borne much more readily than that of saving a statistical life.[60]

In environmental terms, two areas may be mentioned as especially significant. The first is the public assessment of the risks from nuclear power generation, which stays stubbornly at a level much higher than the computed risks published by the industry, and the second is the risk from natural hazards, especially when augmented by apparently irrational human responses like building cities in earthquake zones. As far as atomic energy is concerned, we may note that many of the risks in fact fall into categories two and three of those set out above: the latency period of many cancers associated with radiation makes calculations and correlations of all kinds difficult, and the short life-history of the industry gives a small statistical base from which to predict the future.[61] Difficulties here foreshadow the enormity of the task awaiting analysts who try to apply these ideas to global environmental change.[62]

Chaos

In the world of science emplaced by the eighteenth century in the West, the prevailing model was that of clockwork, and basically immutable. After the discovery this century of the uncertainties of the quantum world, the universe takes on more of the aspect of a cosmic lottery. Stewart describes whole classes of stochastic behaviour in deterministic systems.[63] In these, simple equations can generate motion so complex and so sensitive that it appears random. Simple systems, it appears, do not necessarily possess simple dynamical properties. The weather is one such example: if short-term predictions are strung together to form a long-term forecast then tiny errors build up, accelerating, until total nonsense is the outcome. This leads to the so-called 'butterfly effect' in which a butterfly beating its wings in Tokyo affects the weather in New York. In effect, the variety of possible states of the atmosphere is so great that a return to a state of 100 per cent of the initial conditions need never occur.

Chaos also occurs in ecological systems.[64] The simplest models of population growth can engender periodicity and apparent chaos: some of this may be due to external influence but parts of it seem to be internally generated. Ecology thus may have in it a different dimension from the equilibrium concepts which have underlain so much of its thinking. If chaos is involved, therefore, ecology is a study of disequilibrium. If there were a 'butterfly effect' in ecology, then how could any student of ecology possibly determine what is worth measuring? This throws into doubt any claims that ecology might have had as an over-arching narrative for the whole of human–environment relations: like postmodernism in social and literary theory (see Chapter 6), the foundations are being questioned.

THE CONSTRUCTIONS OF ECOLOGY AND OF SCIENCE

In the penultimate section of this chapter we shall review some of the generalisations that can be made about the ways in which ecological approaches (i.e. those emphasising dynamic interactions between the living and the non-living) and science in general assemble constructions of the environment, and whether these findings can lead directly to a set of prescriptions for human behaviour which will lead to some desirable condition for our species or to some wider framework of consideration.

Ecological approaches

Treated entirely within a scientific framework, a number of general statements can be made as a result of work in the last twenty or so years. The first group is about humanity itself:

- We have the highest biomass of any one species: about 100×10^6 tonnes dry weight or 6×10^{14} kcal of embedded energy. Because of our longevity, the biomass turnover rate is low.
- Our population growth rate (just under 2 per cent p.a.) is high by biological standards for an established population.
- The degree of structural organisation is the highest in the animal kingdom: the exchange of energy, matter, individuals, knowledge and traditions between groups is very marked.
- Of the energy which we use, 90–5 per cent comes from the surplus of earlier ecosystems (i.e. the fossil fuels) and not from active ecosystems of the present day. In terms of the appropriation of the energy of nature, estimates by Diamond suggest that 39 per cent of Net Primary Productivity on the globe is used, diverted or reduced by human activity.[65]

Equally, we can look at the major effects of humans upon ecosystems as producing:

- acute but transient perturbations from which recovery is more or less complete and about which no other useful general statements can be made;
- chronic changes such as permanent land use changes or the continuing extinction of species;
- different energy and nutrient relationships in ecosystems, e.g. by the use of fossil energy for 'subsidies' and by the opening up of nutrient flows, as by long-distance transport of food crops and by accelerated soil erosion; and
- new genetic materials by plant and animal breeding, with even greater potential being conferred by the techniques of genetic engineering.

Putting these together, we can make a number of general statements which result from ecological science about the relations of humans and nature which in effect add up to a construction of the environment:

- Few human effects upon ecosystems are typically or exclusively human except by quantity and combination, i.e. the sheer amount of materials changed, their variety, and their combination in space and time. (But some of the syntheses of organic chemicals, the isolation of plutonium and the splicing of genes might make it defensible to erect a 'humans-only' category.)

- Humans can therefore be treated as normal though highly manipulative (partly through being comparatively unspecialised biologically) members of the components of an ecosystem.
- Virtually all the ecosystems of the planet are now open to widespread exchanges of energy and matter.
- The biosphere is now heterotrophic in the sense that many trophic levels receive influxes of energy capital. Some 8–10 per cent of the second and third trophic levels' energy is from fossil sources.
- There is increasing compartmentalisation of the biosphere into ecosystems of different character and function, as into protected wilderness and park areas, agricultural areas, urban-industrial areas and analogous examples. The simpler systems, and the inert areas, are occupying more space at the expense of mature or 'climax' ecosystems.
- Many natural ecosystems have a high diversity of species, and human activity matches this with a considerable versatility and complexity of organisation which results, *inter alia*, in increasing quantities of energy going through just one species.
- The state of ecosystems throughout the world is now dependent on both the structural complexity and the integrity of human societies.
- There appear to be apparently irreconcilable differences between the long-term stability and maintenance of some ecosystems and the short-term requirements of some human economies.

The last two of these statements are terse formulations of a wide variety of what are generally called *environmental problems* and as such have prompted many ecologists to put forward not merely pragmatic and *ad hoc* solutions, but also to suggest that ecology as a subject leads inevitably to certain values which can form the basis of prescriptions for human behaviour.

Ecology into values?

One of the most famous of these translations is that of the American biologist Aldo Leopold in what he called a 'land ethic', of which the core was the ethical view that the land was not a commodity to be bought and sold but a community of which we humans formed a part.[66] He came to this view through at least two stages. The first was the view that it was the task of the ecologist to find substitutes for the natural order: that is, humanly-managed equilibria. This derived from his early work in game management where he hoped that ecological principles would 'substitute a new and objective equilibrium for any natural one that civilisation might have destroyed'. When he eventually found that this approach had many limitations, he thought instead that nature should wherever possible be left to herself, that doing nothing was better than doing anything. Ecological research might still be pursued because it might bring useful results, in a random manner, from time to time. In the end Leopold abandoned a managerial approach and equated wild things with artistic objects like paintings; the final arbiter of both was to be 'good taste'.

The idea that ecological theory might explain human behaviour in the past (and hence presumably enable us to avoid future difficulties) is at the heart of P. Colinvaux's book on world history: 'We should take the attitude that human social systems eminently ought to be explained by the tenets of population ecology'.[67] In this case it is the idea of a *niche* which is paramount and in particular the niche with all its connotations of access to energy and matter which we learn in our first twenty years. Obviously, the equation is Malthusian since we can only expand a niche in material terms if there are a limited number of us.

These and other strands of thought can lead to outlines for codes of practice related to the necessities produced by ecosystem study, i.e. a new form of determinism. One such is worth relating here, though it has to be couched in the most general of terms:[68]

- the necessity for ecologists to point out environmental pathologies as well as the fact that we now often know enough to avoid them;
- the need for a code to encompass both the long term and the short term (a pail to catch the drips *and* a contract to repair the roof, as it were); integration in time and space of amended behaviour sounds like one of the most difficult things to achieve;
- the necessity for a code to provide guidance for individuals, groups and governments;
- the need for an accent on individual action but also on the imperative for self-interest to be forgone in the interests of the common good, where that common good includes the environment as well;
- a code designed to shift our economic focus from products to processes, i.e. to encompass not only the technical efficiency and the designer label of the water pump but the future supply of water in the well;
- the integration of human activities with the processes of the biosphere at every level from the local to the global. The addiction of governments to territorial sovereignty must be combated by international agreements and by a vigorous NGO presence.

If these tenets were to become the basis for a code then more details would have to be added; two areas stand out as needing attention:

1 the need for audit of the state and changes in state of the planet's ecosystems, including those which are dominated by human activity; and
2 provision of enforceable legal rights for non-human entities such as

forests, lakes and other ecosystems, to give them equality before the decision-making processes (which are at one stage usually enmeshed in legal systems) in the manner of firms or minors, for example.

These statements make it look as if scientific findings are leading directly into sets of ethical and political prescriptions. Yet there is always the danger here that the findings of science will change. In ecology, for example, the classic models of systems tending to equilibrium in an orderly set of successions, with measurable characteristics at each stage, are now being replaced by the dynamics of the patch and the importance of unpredictable, chaos-like succession. Even the longer-term equilibrium about some climatic value is discarded in favour of constant changes, often of an asymmetric nature.[69] So to hang moral and legal systems too closely around the findings and models of empirical science has distinct drawbacks.

In this light, the findings of ecological science and the ways by which it arrives at those results have been reviewed for their ethical content by recent commentators such as A. Brennan[70] and K. Lee.[71] Lee, for example, sets out criteria such as those described above as being necessary for proper relations of humans and environment; then, however, the discussion focuses on the idea of rights to basic human needs, which are taken to be invariant not least of total human numbers. There is also a rejection of values which focus on getting human satisfaction via external material goods rather than asking what sort of person each of us is or could become. Excellent questions, no doubt, but not ones which ecological science can address. Brennan starts out with the idea that no matter of fact can justify a moral judgement and indeed suggests that no natural system has a good of its own in most systems of moral philosophy. But ecology gives us a guide (being one of many frameworks which can contribute and certainly not the only one) to where we ought to be and the need to reconstruct the self in the light of ecological knowledge; in particular we are asked to take seriously the idea that we are part of nature and do not stand apart from it. Each thing is (in part) what it is by virtue of where it is in the natural scheme of things.

Reading off moral or ethical tenets from the natural world is thus a difficult business since there is always the danger of simply projecting on to nature what we wish to believe anyway. The complexities are explored by N. Luhmann in his examination of the modes of operation of human society which are, he says, specifically those of communication.[72] Society cannot communicate *with* its environment, only *about* its environment *within* itself. This communication is, in Luhmann's terms, a 'resonance'. Society now is further differentiated into specialised function systems such as law, politics, economy, religion and education. There are hence levels of resonance within each of these function systems. There can be, as we might expect, too much resonance or too little in any one function system. Because of the differentiation, no one function system can stand for the whole of society and the level of resonance in any one function system does not produce a valid ethic for all. So caution in the formation of moral attitudes towards environmental problems is urged.

The outward connections of ecology can be even more far-reaching. It can be argued that the definitive focus of aliveness is the entire planet: individuals cannot survive except in so far as they are part of some greater whole. Each living individual is like a vortex in a stream: if the flow disappears then so does each vortex. Each and every entity then is determined by its relationships and it cannot be conceived apart from that. The lesson for humanity is that the world is in a sense one's extended body: here is the focus of the world in a particular piece of space and time and the value accorded to the self or ego must be axiomatically extended to the rest. Yet we know that at present this is not so. In western cultures, to privilege the self has become automatic: how to value the rest is now no small problem.[73]

Science and the environment

Does science provide a 'good', or 'successful' or 'viable' construction of environment for us? It has to be said immediately that the main such construction made by present day science and technology is not basically ecological, even though we have devoted a lot of space to that science since its principal phenomena of investigation are environmental. Most science provides for us a much more mechanical view of the world, with direct chains of cause and effect; heirs perhaps to Newton, the world is seen as like a snooker table, whereas the ecological view is that of a mobile in which all the parts adjust to a change in the position of any one, though within constrained limits. One result of this is that the long-range consequences of human-induced alterations of ecological systems are poorly predictable. We may say, *if* we do this then *that* will happen, but we can rarely if ever say when and by how much. But science and its social context (especially that of funding) is highly dependent upon predictive success as an aspect of its role in trying to control the natural world.[74] Open-endedness is not what the public expects of science, for in that case political policies would be erected on a foundation of acknowledged ignorance. One consequence of this for our general attitude to change seems to be that the onus of proof for showing that disturbance is not deleterious is put on the scientist and not on the industrialist or the developer.

In practical terms, nobody can dispute the success of technology in actually building up new systems and structures on the face of the earth, and destroying others in the process. Lest we get too worried about this, we need to be reminded that it is a process as old as humanity itself: there were great man- and woman-made transformations of the face of the earth when fossil fuels were only subsidiary heating sources in remote monasteries.[75] Lately, though, power has been the key. Part of this power is operational

40

in the sense that technology has provided the access to energy sources and conferred subsequent control over them. The other part is the cultural context in which it is acceptable actually to wield this power. This last idea needs to be elaborated a little for, as we have seen, there are those commentators who think that technology provides its own momentum and that *de facto* there is no choice exercisable by humans about its use.

ONGOING QUESTIONS ABOUT SCIENCE AND TECHNOLOGY

Enough has been said to implant the idea that the claims of these activities are regarded by some with suspicion. We need now to rehearse in very general terms these misgivings so that they can be taken up again in the last chapter and placed there in an even wider context.

The outline of scientific method discussed in the first section of this chapter is somewhat idealised. In particular the idea that it is difficult if not impossible to invent a language of observation which is free of pre-existing theory undermines some of the classic scientific positions. This, together with close examination from some twentieth century thinkers, has led at one extreme to a relativism in which theories are basically language-games which are internally consistent in the sense that truth comprises consistency with the particular theoretical system, and knowledge is a socially institutionalised belief. The science may be instrumentally progressive and accumulative but the thrust to some universal truth is blunted if not diverted entirely. Yet it is still a surprise to see articulated such assertions as:

> Science no longer has a single strand, no shared method, no common preoccupation, no values which all its branches share.[76]

Even if we do not go quite so far with relativism, it is worth noting that science has definite limits. Many practitioners agree that the 'hard' or 'aristo-' sciences have that status because they have been able to locate problems that are simple enough to solve, and this excludes most environmental questions. Further, as suggested above, there is perhaps no such thing as *the* scientific method: there is, for example, no naive or innocent message from nature: 'in all observation, we pick and choose, interpret, seek and impose order'.[77] Sense data are taken and not just given: we learn to perceive. The natural sciences, thus, are social knowledge in the sense that an individual's scientific knowledge is made possible by the social conventions of interpretation and by participation in the social process of critical transformation.[78] Science is, in the end, a holistic web of belief which is preserved or reconstructed without being wholly destroyed.[79] The metaphor of the web reminds us that historically western thought has often constructed the world in terms of a single image: in medieval times it was a book, in the Renaissance a human person and in modern times a machine:

41

passive and inert until humanised. So now perhaps a web is more appropriate: it is a metaphor which we shall see used elsewhere by a very different set of thinkers. There seems agreement that human judgement is far more limited than is typically believed: all individuals have a surprisingly restricted capacity to interpret and manage complex information.[80] If it was to overcome this problem that the committee was invented, then there is a human invention of doubtful value.

The ontology and epistemology of technology has raised questions as well. It seems as if technical progress becomes autonomous and hence is self-determinative (like nature) and independent of human intervention; there is a primacy of means over ends and it is all so complex that it becomes impossible to have the good without the bad. Hence the favourable effects are inseparable from, and always accompanied by, pernicious outcomes and in the end technology is said to create more problems than it solves: 'hydra-headed' is a term often used.[81] So technology creates a new intellectual milieu, and all phenomena of economics, politics and other cultural manifestations are situated *within* it. Historically, this was especially so in the imperial missions of Europe in the nineteenth century. The thrust to political and economic dominance was underlain by the assumption that the Europeans possessed the most progressive and advanced civilisation that the world had ever known, and this in turn was underpinned by scientific and technological achievements. More machines (especially arms and railways) and a better understanding of the material world enabled reason to triumph over ignorance and superstition; a Revd J. Cummings said in the 1850s that 'we can upset the whole theology of the Hindoo by predicting an eclipse'.[82] Better still, substitute metal for wood, introduce mechanised motive power, enlarge the scale of an operation, increase the speed of communications and put an unprecedented emphasis on precise timing: no wonder that the railway was the embodiment of a new world. In India again, in 1862, 'thirty miles an hour is fatal to the slow deities of paganism'.

So science and technology as agents of civilisation became a radically new way of looking at the world and of organising societies. By the 1890s, it was asserted that Europeans had the right of access to the resources of backward areas, whose people were unable to make use of them. We might want to ask ourselves to what extent a modified and more implicit form of this type of ideology has underlain much of the 'development' aimed at the Third World in recent decades.

3

SOCIETY AND ENVIRONMENT: MATERIALIST INTERPRETATIONS

In an environmental context, the individual human may be the subject of the natural sciences, as in studies of the effect of extreme temperatures on survival or performance at work. But much more germane at present is the relationship of humanity as a whole to the global environment, and the smaller-scale manifestations of this interaction. Such relationships fall within the boundary-lines of the social sciences and the humanities, and the next two chapters will concentrate on these fields of investigation.

Given the power and authority of the natural sciences, it is no surprise that the interactions of humanity with the environment are regarded by social scientists as fit objects for the kind of disinterested study which is central to the natural sciences. The description and analysis of a society and its environment, for example, can be treated much as if it were that of a population of savannah-dwelling ant-eaters. So far as is possible, the language of description is mathematical and the search is for regularities of a law-like nature and for the building of theory. The terms 'materialist' and 'realist' will be found in the literature.[1] It is not surprising, therefore, that there should have grown up a conventional set of procedures and subsequent categories of knowledge (the equivalent of a 'ruling paradigm' in the natural sciences[2]) in most of the social sciences, to be offset by one or more 'radical' alternatives. Conventional views may of course be very conservative and prefer no future change in any relationships or indeed their holders may prefer to go back to those of the past; they may, however, be interested in reform, usually of a gradual nature. The radicals are more likely to advocate revolution (either intellectually or politically or both), though shades are often found, as with Green politics in West Germany in the 1980s.[3]

As with Chapter 2, most space is devoted to those branches of organised knowledge which seem most important at present and this chapter ends by asking how reliable is the knowledge we have acquired.

ECONOMICS

The heart of the discipline of economics is the way in which it treats the behaviour of humans who find that their possession of means is insufficient

43

to meet their desired ends. Thus economists are, in the world of pragmatic affairs, technicians (like lawyers or dentists) who explain whether the means are available to achieve what we want. The emphasis on means is important, for economics is supposed to be indifferent about ends: they are the subject of the psychologist, the historian and the theologian, for example. Modern economics is, therefore, to some extent different from the Political Economy of the eighteenth century in which it had its origin and in which the disentangling of means and ends (especially when it came to the exercise of power in the distribution systems) was seen as less important. The removal of ends, however, does concentrate within the discipline all the data-gathering and processing techniques of the sciences and their common language of mathematics. Economics can thus strive to become an inter-subjective study on the same basis as natural science.

As we have noted before, economics is currently the major mediator between human societies and their use of their biophysical surroundings.[4] The sub-fields which address themselves most directly to these concerns are environmental economics and resource economics, which are sometimes combined or at any rate not clearly distinguished. Environmental economics in particular sees itself as a way of attributing measurable value (usually as money) to features of environmental importance. Examples might be the direct contribution of natural resources to economic growth (hence the overlap with resource economics), the role of the environment in the quality of life (as in the provision of beautiful places, wildlife or cultural heritage) and the negative connotations of a 'poor' environment which may add to stress and bad health in the human population.[5]

Conventional economics and the environment

By 'conventional', we mean here the western capitalist economics which are found in the free-market economies of the world, as distinct from the centrally-planned ones of socialist nations. The theoretical basis of this view of the world comes from a series of concepts arrayed round the setting of price as the central mediator between supply and demand. The core idea is of consumers and producers, both of whom wish to maximise their satisfaction from a transaction. The first want some form of contentment from the purchase, the second want to make a profit. It is assumed that both are in possession of perfect information about the state of the market, and of sources of alternatives should they exist, and that the supplier does not have a monopoly of the product. There has been, therefore, a tendency to a highly rationalist theory which does not always fit one-to-one with the world of experience.

Environment has sometimes been seen as a set of inconvenient variables for some theoretical models in this type of **neo-classical** economics. There is first of all the question of externalities, which particularly affects our

44

treatment of wastes. A price can be fixed for coal, for example, which reflects the costs of delivering it to the power station. Acid rain from its combustion then falls 1,000 km away in another country and produces a cost there in terms of loss of fish and forests. In a neo-classical economy, these latter costs are of no interest to the supplier and consumer of the coal. In fact, these external costs are subject to government intervention since there is political pressure from the sufferers of acid rain. The government then either subsidises the system by paying for desulphurisation of the stack gases out of taxes, or allows the generating plant to charge more for its product so as to be able to invest in pollution control technology. This latter process is called internalising the costs. Second, many aspects of the environment are not under private ownership and therefore are common to all who can achieve access to them. Under such circumstances it may be gainful for every consumer to make ever greater use of these resources even though this is a source of disbenefit to all the other users. Pollution of the atmosphere or the uncontrolled use of a fishery are examples that come to mind. Thus the whole thrust of the UN Law of the Sea negotiations during the 1970s was to move large areas of oceans and their beds under the aegis of national sovereignty so that they might have an 'owner'.

Economics has been mostly concerned with the environment as resources.[6] These may not always be of the consumable kind, for the discipline has attempted to get to grips with matters such as nature conservation (in trying to put a realistic price on wildlife or trees in the countryside, for example), or on the value of land as scenery or for outdoor recreation. These phenomena, of course, may have some economic value without necessarily being marketable: peace and quiet would be an example. But central to economic concerns have always been the production and consumption of energy, food, metals and other material products of the planet's biophysical composition. These are conventionally divided into (i) stock resources, i.e. those of which there is a physically finite quantity, such as metal ores, and (ii) flow resources, those which are either self-reproducing such as plants and animals, or are naturally cyclic in their movements, like water. The task of economics then becomes to work the price mechanism so as to control the depletion or usage rate of resources. For stock resources, this involves the adjustment of the rate of use so as to encourage the development of a substitute for a resource that will one day run out, i.e. the price must rise so as to produce enough investment for the research and development of a new product and to encourage consumers to move over to the new material. The substitution of plastics for metals in recent years can be seen as one example of this. In the case of flow resources, price becomes a regulator of flow which must adjust demand to the capacity of the system for renewal: to the flow of water in a river at its lowest rate, for instance, or to the reproductive capacity of a fish population.

These largely practical considerations are underlain by a great deal of discussion about the relations of price and value. Argument takes place about what should really constitute the value of a product which is eventually reflected in its price. One school of thought holds to the **consumer theory of value**, in which the paramount concern is simply what people are willing to pay for e.g. pearls or wilderness. The high price of pearls is not then determined by the cost of extracting them from unyielding molluscs but by the apparently bottomless purses of matronly ladies. So the value of something is the highest price that consumers will pay in order to have it. This is, *inter alia*, the economics of the brothel and does not always fit the concerns for the environment very well. For example, how can the consumer determine how much to pay for an absence of excess carbon dioxide in the atmosphere? In practice, decision-making here is more political (in response to pressure from scientists) than informed by economic thinking.

In some contrast, there is the **social theory of value** in which it is the needs of the community that constitute the source of value and not the willingness of some consumers to pay an inflated price because they are rich. Thus the provision of a supply of clean water ought to reflect the absolute need of every member of the community for this product and be priced accordingly, i.e. at the lowest possible, so that nobody is cut off from it. A difficulty here is that it may then be difficult to manage a resource so as to fulfil all the demands then placed upon it. The contribution of the actual workers themselves is dominant in Marx's **labour theory of value**. Goods are only worth the cost of the labour which has gone into them and the materials of nature are not worth anything in themselves. Indeed, Marx made remarks about the overcoming of nature which echo this theory and so has been thought to be against those who espouse some form of resource thrift and general environmental tenderness. A radical departure from most of these socially-centred types of thought is made by B. Hanlon who suggests that energy supply is so central to all our activities that **energy** itself could form the underframe of a theory of value, and indeed determine the cost of goods and services on the basis of the embedded energy in them.[7]

The techniques employed by economists to help with decision-making include cost-benefit analysis (CBA). This is primarily a technique designed to improve public decision-making, by assessing the benefits and costs of a project to all groups involved, including intangible effects providing a price can be put on them. A CBA has to consider which costs and benefits are to be included, how they are to be calculated, what the relevant interest rate for future discounting is,[8] and whether there are administrative and political constraints. Despite the apparent objectivity of the process, there have been many complaints about CBA. If it is applied to a series of independent projects, then it may lead to conflicts between each one or between other desirable ends.

46

The types of economic values that can be used in CBA are:

- User values which derive from the current prices which are paid for access to them.
- Option values. These measure a willingness to pay to forestall some future probability, e.g. a payment to protect a landscape from development for a particular number of years or 'indefinitely'.
- Intrinsic values which derive simply from the existence of e.g. a biological species. Such values derive from concepts of altruism ('they have as much right to be here as us') or stewardship ('it is our duty to ensure their perpetuation').

In thoughtful discussions[9] of the true nature of CBA, these are placed in the context of such factors as:

- Irreversibility. In the event of an undesirable outcome, is it possible to reinstate former conditions? A cleared slope can be re-afforested but can a runway be un-concreted?
- Uncertainty. The future always holds risks of an unspecified nature. Can any model build in a capacity to compensate for unforeseen effects of a given project?
- Uniqueness. Both organic evolution and cultural development have thrown up unique phenomena, like a given species or a monument like Angkor Wat. Should this confer on them a very special economic value?

The actual measurement of the value of environmental phenomena has been the subject of much exploration in the 1980s, excellently summarised by Pearce et al.[10] Their contribution has been especially to collect the various parameters needed for a proper valuation of environment which is not detached entirely from ethical concerns. They are concerned with:

- The value of the environment (natural, built and cultural) in promoting both materials, services and a contribution to the 'quality of life'.
- Futurity, both in the short-to-medium term of 5–10 years and in the longer-run future beyond that.
- Equity, in placing emphasis upon the disadvantaged members of human societies ('intragenerational equity') and on not closing off opportunities for future generations ('intergenerational equity').

These items are the everyday stuff of neo-classical economics but are now placed, as a package, where every economic evaluation must consider them. Future generations must be compensated for reductions in the endowments of resources brought about by the actions of present societies. In particular, Pearce is adamant about the proper accounting of externalities.[11] In the past it has often been possible to measure the benefits and 'efficiency' of a development project without regard to all the negative spin-offs. CBA has no doubt been manipulated to justify decisions made on other grounds (it is not difficult to specify the values of the inputs once it is decided what the result is to be) and it cannot measure a great number of the things upon which human happiness depends.[12]

Only if the measures described above are adopted can we begin to talk of sustainability. Mostly out of concern for the Third World, but applicable also to the First, economists have attempted a deeper analysis of the notion of sustainability. The concept is backed by essentially ecological notions of stability and equilibrium which may not be isomeric with socio-economic activity, and its economic analysis exhibits this history. Thus it consists of:

- resource harvest levels no higher than regeneration rates; and
- the input of wastes no faster than the receiving systems can assimilate them.[13]

In more conventional economic language, then:

- sustainable economic growth means that real GNP per capita is increasing over time, and that the increase is not being undermined by negative biophysical impacts or social disruption; and
- sustainable development means that per capita well-being is increasing over time, subject to the same feedback constraints as in the case of economic growth.[14]

Thus future generations should inherit a supply of wealth (artificial and natural) no less than the stock inherited by the previous generation, and their natural assets should be no less than those of the previous generation. One problem with sustainability in economic terms is that its essentially biological inspiration may not fit well with an economics whose stimulus was of a more mechanical and atomistic nature. Another is that it can be an elusive goal that provides a convenient excuse for an endless epistemological search while the forests burn and the estuaries choke. More deeply still, it is possible that the world, at both evolutionary and ecological time-scales, is far from equilibrium and so the biological underpinnings were themselves based on a mistaken interpretation of the nature of the world.

The argument that capitalist economics is basically a tool of the rich within a society has led to the continued attractions of socialist economics, and the movement to try to assert the primacy of the findings of ecological science over the dictates of economics in public policy has led to advocacy of alternative forms of economics, some of which are less determinedly western in their world-views. But both operate for the moment within a global context in which free-market economics is dominant[15] and indeed gaining ground.

Socialist economics

If classical free-market economics is founded upon the self-interest of the individual, then socialist economics has a starting-point in a view of society as a whole. There is an analogy with the natural sciences here and indeed the fathers of European socialism saw their work as being the equivalent for human history and social development of Darwin's theory of evolution. These founders were Karl Marx (1818–83) and Friedrich Engels (1820–95), whose work was fundamental to the school of thought called Marxism. Not only do the writings of these parents bear continued exegesis, but numerous disciples and intellectual descendants have developed their ideas, so that the simple term Marxism is shorthand for a complex set of evolved constructions. Versions of Marxist constructions are also the basic socio-economic structures of the world's remaining socialist republics.

The underpinning is an explanation of the operation of capitalist societies, with several interlocking models providing the initial keys.[16] Of these the most important are:

- The analysis of the circulation and accumulation of capital.
- The social organisation of that type of economy and the way in which this leads to the exploitation of one social class by another and to the degrading of natural resources.
- The operation of the ideological apparatus, including, for example, science and neo-classic economics, in order to consolidate the position of the ruling class.

In our present context, the second section is of most interest, although the last model opens another perspective on our earlier discussion of science and technology.

Critics of Marx and Engels have argued that Marxism is especially hostile to the environment, regarding it largely as something to be conquered and put to use as a resource. In particular, the labour theory of value assigns no intrinsic worth to nature: it is only the human transformation of it which counts in this model. One recent interpreter wrote that, 'The new society is to benefit man alone, and there is no doubt that this is to be at the expense of external nature. Nature is to be mastered with gigantic technological aids'.[17] Thus anti-Marxists level two main charges:

1 That Marxism pits humans against nature, in precisely the way that modern capitalism harnesses technology simply to cater to immediate material demands.
2 That Marxism denies any value to external nature.[18]

Overall, there is perhaps a trend from a more environmentally sensitive position in Engels and early Marx to a harder line in the later writings of Marx. Engels was particularly strong on the long-term ecological effects of human actions, and Marx warned that capitalist agriculture was bound to lead to loss of soil fertility. He also inveighed against the idea that land had become a mere commodity and thus passed from one usage to another without regard to its historical and natural characteristics.[19] So if we accept that the earlier discussion still underpins the later views, then a case can be made for Marxism as being quite sophisticated in its view of nature. In particular, Marx did not want a return to some pastoral Arcadia with its connotations of zero economic growth for he felt that nature was full of untapped potential which was held back from people by the chains of the class structure.

It can be further argued that Marxism is in favour of an open-ended dynamism in which both man and nature are subject to constant transformation; this process affirms both parties in their dialectical opposition and unity. Marx in fact spoke of nature as man's body.[20] The effect of the industrialisation of the nineteenth century made him see that nature did not then exist in entire independence of human actions and values. The logic of this interdependence is to maintain and strengthen relations between the two parties: not only to satisfy material needs but those of the mind also; both Marx and Engels thought that aesthetic nourishment was important.[21] Again, what prevented this desirable end was the set of social structures then present which kept power in the hands of those who controlled capital: disharmony between man and man was bound to lead to an equivalent between man and nature.[22] This dichotomous view has prevailed to the present decade: socialists will argue that capitalism is more environmentally degrading since production levels are not so great in socialist economies.[23]

Integrating ecology and economics

Bringing together the epistemologies of ecology and economics is not easy. There is, to begin with, the deep suspicion with which environmentalists regard economists, often thinking of them as the hired hands who provide the justification for much environmental destruction. Getting beyond that, there are some intractable intellectual problems, fundamental to which is the development of a common metric.[24] So-called 'marginal opportunity costs' have been one obvious route, but in general have fallen under the type of suspicion that has befallen CBA: you put in the numbers that will give you the right result. The most popular common denominator at present seems to be energy[25] and others have revolved around game theory, energy-plus-matter and even evolutionary theory.[26]

Since energy flows through an ecosystem and also through an economic system, it is not surprising that there have been attempts to use it as a common measurement to link both systems, with the possibility of using the current prices of commercial energy to cross-value the energy present in nature. We can thus put a price on e.g. sunlight, which is not simply plucked out of the air, so to speak. The actual methods of performing the calculations involved in the cross-system valuation of energy at various scales such as the region and the nation are too complex for this book but are readily accessible. The key measure for economic systems is usually

embedded energy, with the alternative name of **embodied energy**. One complexity is that ecological systems usually run off sunlight, and economic systems off fossil fuels. Yet the calories from these two sources are not directly comparable because they are of different qualities, are suited to different needs and capable of doing different amounts of work. Hence it is not necessarily a direct calculation to see how many calories of sunlight are embodied in the production of a book, for instance. But if we accept the assumption that all products of the human system, including labour, ideas and information, are evolved from natural systems (including fossil fuels), then we have a basis for comparability and cross-calculation; the work of M. J. Lavine shows, however, that this is not yet a simple matter for the pocket calculator.[27] However, it has sometimes proved an irresistible conclusion that if energy input is greater than energy output then the process must be undesirable. Looked at cosmically, such a view would preclude life altogether, since low entropy is consumed in building up the complexities which make life possible. Noting, however, studies which purport to show the not-too-distant exhaustion of fossil fuels or mineral ores, or a coming crunch between growing human populations and food supplies, some economists have tried to construct an economics of scarcity rather than of expected abundance. The distributional function of economics is thus intensified: the scarcity of resources is likely to be a matter of life or death (or at the very least of life-style in the industrialised nations), not simply of the price of petrol. Economic adjustments mostly centre upon the role of governments in replacing the mechanism of market price with something more appropriate to an age of exponential rises in use rates of materials, in anticipation of the '29th day effect' which, it is said, is not responsive enough to market mechanisms.[28] This type of economics has been especially influenced by the work of K. E. Boulding, who used the metaphor 'space-ship earth' to highlight the fact that we live in a limited sphere; he contrasted this view of our economic situation with 'cowboy economics' which emphasised the profligate burning of cheap candles at both ends.[29] Boulding has also tried to marry the basic thrust of evolutionary theory with the construction of human economies by treating human history as the evolution of humanly-fabricated artefacts.[30]

One of the ensuing complexities is the realisation that like any study of human-environment relations, the interface of ecology and economics has of necessity to recognise the existence and relevance of matters as diverse as the laws of physics (e.g. the second law of thermodynamics), law, individual psychology, sociology (e.g. the social aspects of carrying capacity) and spatial array of the kind studied by Human Geography.[31] So there is not yet an integrated construction resulting from this particular interfacing. It has, however, succeeded in planting rather more firmly into economics the notions of limits in biophysical systems, or, put more vulgarly, there is still no such thing as a free lunch.

51

Other alternative economics

Marxism is not the only alternative system to the political economy of capitalism. Writers have developed other sets of ideas, although it has to be said that, unlike Marxism and capitalism, they are nowhere in place on the national scale for inspection and evaluation. If we accept that the disentangling of ends and means is very difficult where the human use of commodities is concerned, then it is not surprising that non-western value systems have been examined for their potential contributions to alternative economics. For instance, as much out of traditional values as out of today's context, E. F. Schumacher brought to our attention his conception of 'Buddhist economics'.[32] This combines the Buddhist ethic of detachment from the pleasures and pains of the world with the attitudes of J. K. Galbraith towards the new corporate state. They meet in an exaltation of the frugal and thrifty, small-scale living and decision-making, and a renewed attention to the needs (as distinct from the demands) of individual people rather than corporate stockholders. This type of economics has appealed to groups of people seeking alternative life-styles, often on a communal basis, but not much to governments which do not fancy isolation: in a world of wolves, do not lose your teeth, runs the proverb.[33]

In a more directly western context, the case for alternative economics rests on two major foundations:

1 The market mechanism for allocating resources through time is inefficient. Equity and uncertainty considerations are at the heart of such problems and the answers are manifold in both the technical sense (e.g. different kinds of discount rates) and the political arena: what is a feasible time-horizon for a decision-making body? These views emphasise that 'efficiency' is only one thing to be optimised, along with e.g. population growth rates, decentralised settlements and the nature of income distribution.
2 Marxists and other advocates of an epistemology based on socio-cultural constructions of environment need on the one hand to recognise the realist finding that there are indeed physical laws, and on the other that technological change of itself will not necessarily alter relationships such as poverty or the desire to change nature irreversibly.[34]

Indeed, both lineaments may focus on the resurrection of the idea of an **absolute** natural scarcity, sometimes buried by theorists of classical and neoclassical persuasions.[35] This time, the notion could encompass not merely stock resources that were no longer available for extraction, but widespread environmental degradation and ecosystem collapse, as with the worst forms of desertification, for example. Alternative economics then begins to advocate the efficiency and value of factor substitution (labour for energy, capital for energy and materials, increased efficiency of resource conversion) and improved organising techniques which encourage the change to durables. The 'throwaway' economy is replaced by long-lived products where design builds in not only long life *per se* but proper possibilities for repair and recycling. This is consistent with thermodynamic analyses, where the real savings are made in longevity of the product.

Summary

What is the future for economics? Although it seems to preside over unemployment, inflation, capital scarcity for the poor and environmentally destructive notions of 'development', the immediate future seems very much like business as usual. Though some argue that the whole of western society, economics and all, is in a phase change to another type of relationship with the non-human components of the planet, there seems little evidence of this in the financial pages of our daily paper or of the *Wall Street Journal*. Perhaps economics is in a sense epiphenomenal to the rest of society, i.e. it merely operationalises (like a good technician: see the opening paragraph of this section) the values of society towards its goods and services, and therefore lags in any phase change rather than leads. In any case we seem unlikely to see it dethroned from its present position for some time yet: the most potent construction of the use of the environment for humans will still be 'is it economic?'. In this case, environmentalists will do well to understand what economics is about, particularly welfare economics, and the light which the discipline sheds upon the complexities of getting a living, for example. The suspicion has grown that much environmental destruction has been judged profitable by hired economists and that they must therefore be regarded as embodiments of some Manichean evil: on the contrary, every encouragement should be given to them when they show signs of getting greener, a trend which can be seen in some of the larger organisations such as the World Bank.

The tasks awaiting economists appear to be threefold:

1 The interfacing of the ecological systems of the world with its economic systems. Neo-classical economics has always assumed that the economy is always below biophysical limits, which seems now to be turning out not to be true unless a spectacular 'technical fix' for some problems of residuals like the 'greenhouse gases' is discovered. Feedback loops between ecological and economic systems must be a subject for special examination.[36]

2 The exploration of the bases of consumer demand. J. S. Mill's famous dictum, 'Men do not desire to be rich, but to be richer than other men' no doubt has its truth. But there are yet stronger pressures in the higher-intensity market-setting of industrial societies where individuals can be led to misinterpret the nature of their needs and the ways in which they

can be satisfied. There seems to be no sense of contentment nor well-being in the higher reaches of abundance.[37] Somehow the self has to be redefined, which is a task at which even economists baulk.

3 Some economists see that the metaphors of the approach need careful thought. Economics appears to have taken on the atomistic and mechanical assumptions of classical mechanics of the type developed by Isaac Newton. Yet economics interfaces with an evolving set of interconnections of a probabilistic type, in the shape of the planet's biophysical systems. In particular, the adoption of seventeenth-century mathematics as a basic language has, according to K. E. Boulding, set economics on a path that increasingly has become a dead end.[38] The number of propositions about social systems which can be reached only with the aid of mathematics is, he says, quite small. Mathematics is deficient in verbs, so there are limits to what can be talked about: it favours structures over process and equilibrium over evolution.

Boulding's assessment of econometrics suggests the need for economics not to forget its roots in political economy. Here the question of politics is not left out, so neither is that of the exercise of power and hence of values. Dignity, freedom and 'happiness' are not guaranteed by economic efficiency any more than by anything else[39] though we can gain some comfort from Boulding's view that in the end economics has done more good than harm.[40]

POLITICAL SCIENCE

Politics is the deliberate and rational effort to direct the collective affairs of human beings. In that pursuit, then, politics is about power, and the ways in which that is conferred on individuals and groups and what they then do with it. In the case of the environment, political science also aims at an objective, value-free study of what actually happens in terms of decision-making as to the allocation of resources. In the sense that it tries to find optimal structures or organisations to accomplish such tasks, it is operational in scope. Other parts of the discipline, such as political philosophy, are more academic in tone, though they may set the agenda to create an atmosphere of acceptance for particular types of structure.[41] We shall, however, consider political **movements** like Green parties in the section below on sociology.

The classical tradition

The agenda for environmental politics was in some ways set in the time of Plato and Aristotle, though admittedly in a largely pre-technological context.[42] They and their contemporaries wrote, among other things, about population levels. This led to concern about how the numbers of people interacted with natural resources and also how population levels affected personal psychology and hence the imperatives of social organisation, the *politeia*. Likewise 'nature' was of considerable interest, in both its 'natural' form and as 'human nature'. For both, the questions could be asked, 'what is permanent and what is changeable?' and similarly, 'what is the proper attitude to sudden eruptions of instability, whether tectonic or political?'. And in particular, perhaps, 'are the desires of human nature insatiable?'. How are they affected by the distribution of wealth or by human closeness to or remoteness from the natural world?

The spirit of the classical tradition was in pointing a direction rather than laying down an exact prescription. Then, as now, two distinct strands of thought emerged. There is first the radical, indeed revolutionary, view that wishes to overturn historical tendencies in the name of humanity's best qualities (and in the belief that planning ahead is essential) and in particular not to subjugate certain human qualities to economic and technological imperatives. This can result in a position well to the 'left' of European socialism today. Second, there is the conservative outlook which looks for a society in which change is limited and preferably subject to authority; such an organisation contains a tension between social and technological orders which are unexamined and uncriticised by tradition and the desire to let individuals choose freely among competing alternatives. *Mutatis mutandis*, things are not so very different today.

Conventional politics today

The normal political activity which concerns us here is that of the formulation of environmental policy. This can take place at many levels, from that of a local government deciding where to put its dump for householders to leave their rubbish to the actions of UNEP in the face of desertification or the apparent problem of global warming. The management of resources and their residuals, as well as the designation of land and water for non-consumptive use, is always a political matter as well as being economic and scientific.[43] In general terms, we can contrast societies with openness of information and decision-making with those which keep as many secrets as possible and involve only the bare minimum of power-holders in the exercise. The USA is a good example of the first category, the UK and France of the second. The nations of the South are different again, for they mostly lack the institutionalised structures to enable them to tackle the complex problems that have a large environmental component. The UN can only act as mediator and catalyst: where others are not willing then nothing happens, as shown by programmes like UNCTAD and UNCLOS when they attempted to give the South a better deal. The refusal of the USA to sign the biodiversity treaty in Rio (1992) is another example of industrial-world hegemony.

The distribution of power in more favoured places is also subject to review by political scientists, who see the nation state as an ineffective manager of environment, for it is often smaller than the ecosystems over which it seeks control. Here a possible solution is to invest each individual person with political and economic rights and then to try and protect the health (mental and physical) of all at community level, so that a relatively small group can acquire a balanced and secure economy.[44] At present, we seem usually to have an overwhelming central government against which the only countervailing force is the presence of interest groups, of which there are many with environmental concerns, like Friends of the Earth and Greenpeace, as well as those campaigning for a specific species, issue or place.[45] These groups may achieve some measure of success since participation in the processes of persuasion makes them highly professional and technically capable. The price of acceptance by the establishment is usually 'responsible' behaviour, which does not always appeal to their active members because their influence is usually less than that of economic-interest groups like the Trade Unions. Such groups try also to alert people who are otherwise sometimes willing to accept short-run 'solutions' which may eventually deprive them of their liberties: fear of energy shortages has led some European nations into nuclear power programmes which threaten not only with radioactive wastes but with authoritarian structures in order to maintain integrity of the unclean fuel cycle.[46] A counter-trend argues that environmentalists' views could only be emplaced by highly authoritarian governments and so charges of 'eco-fascism' are levelled, with the implication that democratic change towards such different world-views is impossible.[47]

Alternative politics

The land, industry and rapid urbanisation have for long been a fertile seed-bed for political alternatives, from pantheistic mysticism to varieties of socialism.[48] The anti-growth movement of the 1960s and 1970s was primarily aimed at economics, but there was spin-off criticism of political science as being merely an agent of the *status quo*. This led to the development of a politics of the steady state, which was usually so strongly entwined with economics as to revive the former framework of Political Economy. In this context politics becomes not only the art of the possible in the sense of working within existing paradigms but also of creating new possibilities for human progress. Steady state politics was concerned with the ways in which a less flamboyant life-style could be accepted by the rich so that the poor might become less deprived.[49] Again, community was to be the bedrock of decision-making, informed by ideas of design rather than planning, to allow more individual freedom, and by notions of stewardship rather than the primacy of consumption. The community-based approach is founded on a reaction to the globalisation of economies and politics. It sees a healthy humanity–nature relationship based on a local ecology (shades of the *pays* of the French geographers of the inter-war period) which refuses to co-operate with oppressive institutions from outside the local economy. The 'bioregion' is seen as a site for political action, cultural and spiritual expression, and personal change, all enhancing an ecologically-based trans-formation of self and society.[50] No doubt many individual humans enjoy being rooted in a 'local' community but for others movement, change and anonymity may be essential for creativity.[51] And how could thousands of 'bioregions' mesh together to provide higher-level services like the more hi-tech medical facilities?[52]

Political change

The desire to change political structures for the better is subject to illusions. Boulding has pointed out a number of these.[53] The first is the perception that the rich take away resources from the poor whereas the poor produce little anyway and the problem is one of differential development and access through time to knowledge and capital, combined with effective organisa-tion. The second is founded on class lines and avers that social class is identical (when it comes to asserting control) with an interest group. But neither workers nor capitalists of the world ever in fact unite: they are always divided by something else. This leads to the third illusion: that revolution will put the virtuous in power. But a good revolutionary is often a bad governor and in any case revolutions seem to produce stratified societies like those they replaced. A milder version simply says that if every source of oppression is removed then all will be well.

Yet it is a fact that startling changes do occur in political structures: the abolition of slavery is an example as are the changes in Eastern Europe and the USSR in the late 1980s. It seems that such transformations come about through a wholesale process of transmutation of world-view, a *metanoia*.[54] Yet it normally takes decades for the necessary shifts to occur in enough of the population; if we believe the parable of the twenty-ninth day, then it is open to us to wonder if we have time enough to spare for such leisurely changes. By way of a current example, we should keep track of the reaction of the world community to the scientific models of global warming. Immediately, there are two major divisions: those who prefer not to believe the science because it is not 'proven', and those who believe that it is better to be prudent, in case it turns out to be right. Within the latter, there are liberals with faith in a free market approach to curtail e.g. carbon emissions to the atmosphere, and regulators who believe that legislation and taxation (like carbon taxes) are essential. Beyond this there may, or may not, be the same kind of conversion experiences that ended slavery: environmental questions too have a habit of turning into questions about justice as well. One interesting development of the 1990s has been the emergence of **political ecology**, a way of approaching environmental concerns which places humans acting politically in the foreground. Here, the question of the distribution and exercise of power is explored, in both historical and current contexts.[55] In particular, this insight can be applied to an agenda embracing, for example, environmental reformism (i.e. 'enlightened capitalism'), radical community movements, the Third World, ecofeminism and the reconstruction of Marxism.[56]

Politics in a wider context

The relations of technology and politics have already had a brief mention. This interaction has always been a subject for comment by political theorists from Classical times onwards, with more than one Utopianist calling for an authoritarian body to oversee the introduction of new inventions into society, with a strong presumption that a conservative position was preferable. There has been little chance of that happening since the nineteenth century, and even less today at a public level: only the original inventor can now keep a secret and in any case he or she is likely to be a member of a team.

There are direct instances of the effect of technology upon social order. The low bridges on the Long Island (NY) parkway were meant to keep out buses and so keep out blacks. Hausmann's boulevards in Paris made barricade-building very difficult. The invention in California of the tomato harvester produced a decline in the number of growers from 4,000 to 600 in the period 1960–73 and by the late 1970s job losses in that industry totalled about 32,000. So technologies can be not merely symbols of a social order which contains power structures but more literal embodiments of that order.[57] Technologies thus are ways of building order: choices made by people with power are strongly reinforced by investment patterns, materials and equipment and social habit. To be opposed to these patterns is generally seen as being not merely anti-technological but anti-progress as well.

Given such formations we might expect indirect effects of technology upon political patterns as well. From Plato onwards, the question has been asked: if you introduce an invention then is a politics of freedom or of repression part of the package? For example, a sailing ship at sea needs an authoritative captain; a railway system needs a hierarchy of spatial control over a large area. By contrast, it is said that small-scale alternative energy technologies bring choice to individuals and small communities. Included in this comes the power to determine a slice of their own futures: once again, access to energy sources confers political and social power. This is to over-simplify, of course. Some technologies are well adapted to a centralised and hierarchical authority, like the manufacture of motor cars on an assembly line. Yet it is possible to make cars (at a profit) using more autonomous work teams. At the extreme, though, there is an economy (not yet with us) highly dependent upon the use of a lot of plutonium. Then its theft would be loaded with such serious consequences that many civil liberties would have to be forgone (first by the plant workers, then by the rest of society) in order to keep track of every last gram. Anybody who opposed such checks and controls would be dismissed as having too little care for their fellows.

The main danger, perhaps, is that the model of factory or firm is taken as the epitome for society as a whole. That is, if a firm makes most profit by being structured in an hierarchical and highly authoritarian way, why should not the same apply to society and, by extension, to our use of resources and hence our manipulation of nature? This is the theme of much of the work of Murray Bookchin,[58] who postulates a hidden prehistory of the development of a landscape of domination by and within human hierarchies. Not only is this expressed in our visual world of experience but seems to be embedded in the depths of our psychic apparatus. Part of this has involved our disembodiment from an 'external' nature which some now see as mineralised and inorganic. The alternative (which is usually incorporated with Green politics, see p. 62) is more mutualism and self-organisation, with subjectivity of experience having as much validity as 'objectivity', spontaneity and non-hierarchical relationships. The language, at any rate, is reminiscent of the theory of evolution and of the entropic theories of organic organisation. Thus, alternative models for human politics do exist, though routes from here to there are as always full of barriers, potholes and low bridges.

SOCIOLOGY

Sociology undertakes to observe and describe social phenomena and to formulate theory in a scientific way. As it has developed, it has addressed itself to questions such as, how does a society hang together, i.e. what are the forces of its 'social physics'? The analogy with the physical sciences and, in particular, a Newtonian view of the relationships between entities is apparent here. It has been interested as well in social evolution; in the main this has been underpinned by a progressive view of history, though a few pessimists have always surfaced in any generation.

It cannot be said that environmental constructions *per se* have been a major preoccupation of most sociologists in the post-1945 years but that their lineaments of thought could relate to such a field of enquiry need, however, not be doubted. In terms of social forces, we might ask what individuals and groups within society espouse the environment as a major cause and why; in evolutionary language, we might well be interested in whether a society is structured so as to be well adapted to the environmental constraints it experiences and whether it recognises that they exist. The facts of life in a social context mean that there are always people wanting to change the structures of society, either in the short or medium term through political action or in the longer term via the creation of Utopias.

Mainstream sociological enquiry

Sociologists are the authors of numerous reports on peoples' attitudes to various environmental issues, from the purely local (do you want that meadow covered with houses?), to the national (are young people sufficiently interested in the environment?) and the global (are you willing to roll on your deodorant in order to protect the ozone layer?), and many more. The work of S. Cotgrove goes beyond this in identifying two social groups with particular environmental attitudes.[59] These he calls the Cornucopians and the Catastrophists. The first put their faith in technology and economic development and assert that increased quantities of resources can easily be available for all provided that investment in technology is high and that social structures encourage enterprise. These are the dominant group in the world at the moment. The second group thinks that there are physical limits to resources and that the planet's life-support systems can be badly degraded by environmental contamination; reform needs attention to wastes and to a lower level of material consumption in the industrialised nations. The viewpoint of the sociologist is that all these views of the future are rooted in systems of meaning which are themselves social constructs and lack any basis of objective certainty. They are to be identified with faiths and doctrines just as much as any religion. (Cotgrove adds that since the Cornucopians are dominant, their views need the most stringent examination.)

In the critique of the Cornucopian position, the lead has been taken by the 'limits to (physical) growth' school of the Club of Rome and its associates.[60] A notable study, however, parallels this with the idea of social limits to growth. In it, F. Hirsch[61] concluded that economic liberalism has its limitations in that it promises prosperity to all, provided they obey the rules of the game, but in fact unleashes demands and processes that no western society (nor, we suppose, those of the South) can contain. Because so many goods lose in comparative quality if they are open to all (those of the environment like open space or a second home are good examples), keeping ahead of the pack is still important for many: only that 'undiscovered' island in Greece will do for holidays. Thus most people never improve their position relative to others and the resources experience increasing pressure.

Radical social theory

This side of Utopia, visions of societies which are in a greater position of harmony within themselves and with nature have their attractions. As a counterweight to the industrialised state of the late nineteenth and twentieth centuries, a process of change towards a more decentralised and de-urbanised set of communities is advocated by many writers. Such communities are generally experiments in communitarianism and controlled anarchism, and were perhaps more popular in the 1960s and 1970s than today. Nevertheless, as T. Roszak has expressed it,[62] it seems more fruitful to seek out hidden well-springs than to be resigned to a future with only waste lands. Roszak has gone on to encompass the Gaia hypothesis within his social thinking and to suggest that Gaia can communicate directly with her citizens and thus adapt their behaviour so as to be life-enhancing in the long term.[63] The mode of reception of this knowledge is what we call intuitive.

Whereas some social commentators see gradual change as the only way in which a true 'greening' of society will occur, others follow Marx and Engels in foreseeing revolution. This is the position of Herbert Marcuse, who sees the necessity for a technology of liberation which will succeed the present technology of repression, and that the one must succeed the other very rapidly, i.e. at a revolutionary rather than an evolutionary pace.[64] The revolution will, he sees, be brought about by the world's underclasses and be one in which social and political judgements are explicitly made about technology (i.e. some forms of research and development would be explicitly forbidden). The role of nature in the new world is central to Marcuse. He wishes to see its rediscovery as an ally in the struggle against exploitative societies in which the violation of nature aggravates the violation of one human by another. He sees that nature in its controlled state has become another instrument for the domination of other people and that it should instead be recognised as a parallel life-force, a striving for a diverse and enhancing life that is characteristic of human societies as well. Since nature can also be seen as a cosmos with its own potentials and necessities then it can also be the bearer of objective values. Marcuse thus directs us towards a set of processes of change, though he does not specify the exact type of society which would result.

Bookchin, on the other hand, identifies the problem in a much more specific way.[65] The core is seen as hierarchy in human societies: thus the pyramid of dominance enables us to negate so much potential for liberty and freedom. In such a structure, the ethical meaning of the community has been replaced with what is termed the 'fetishisation of needs', i.e. a society devoted to satisfying, at whatever cost, the demand for consumer goods. In such a situation, even the state of nature becomes a commodity, witness the bookings needed to get into Wilderness Areas in North America. As far as the environment is concerned, Bookchin's central message is that we must stop dominating each other if we are to stop wreaking havoc on nature. Nature becomes, in his construction, something that can be dominated by those who themselves are dominated by others and is thus at the bottom of the hierarchical pile, with women somewhere in the middle.

Green politics

It seems more appropriate to include this topic here as a social movement growing out of the above discussions than as a coldly observed piece of political science.[66] Green political parties are devoted to moving towards the Utopias of the environmentalist cause (and are Catastrophists in Cotgrove's terminology), and in particular to a low-consuming economy, to the use of renewable energies, to decentralisation, to the re-evaluation of the roles of e.g. work and women, and especially to the peace campaigns which aim at the removal of nuclear weapons.[67] Since the end of the Cold War they have also turned their focus on to other conflicts, especially those involving environmental damage like the Gulf War of 1991. In countries with sophisticated voting systems (notably the former Federal German Republic), the Greens have done reasonably well at most levels of representation; where first-past-the-post applies, then they have a great struggle. They are, however, different from other political parties in the sense that they wish to put themselves out of business, having converted all the other parties to their world-view. In general, though, they are not anarchists: they advocate participation in a structured society, albeit one devoted to different ends from those of the present in the industrialised nations. Many see it as inevitable that conditions will get worse before the mass of people are converted to their way of thinking.

Strains within the Green movement are, however, often evident. These are usually between the purists who want to move directly to the new consciousness and values of the Deep Ecology (see p. 135) type, and those of a more 'shallow ecology' type who will join coalitions with those who accept the prevailing patterns of production and consumption. These latter are engaged in a form of social engineering which aims to bring about better short-term conditions for humans rather than a change in our whole world-view. They do, nevertheless, wish to lessen our impact upon nature and to reduce contamination levels.

HUMAN GEOGRAPHY

Geography can be explained as the branch of knowledge which describes, classifies and explains the distribution of material and human phenomena in the space accessible to humans and to the activities which they control. Since it brings together both the physical and cultural worlds, it might be thought of as *the* discipline of environmental constructions but its plurality of approaches and its recent history have combined to prevent any such development.[68] Increasingly, human geographers have moved away from any taint of environmental determinism and so they have been less and less at the interface with physical geography than ever before; hence the thrust of many human geographers to place themselves securely within the contemporary boundaries of social science.[69] This has entailed the adoption of both positivist methodologies and more reflexive attitudes.

Orthodox human geography

Within the present context, the main trend to note is the strong move towards a positivist geography in the period after 1950. Much of the work before then was characterised as dealing only with unique instances and the later work, which tried to use mathematics whenever possible, sought to bring human geography into the positivist fold by seeking generalities that could be described statistically. Laws and theories with as much validity as any others in the social sciences were sought and indeed it was the aim of some geographers to get up alongside the physicists; gravity models of commodity flow between places were seen as steps along that road.

The founding documents in this movement were seen as the studies in locational analysis which had pointed out regularities in the spatial outcomes of various economic activities. The most important of these, probably, was the nineteenth-century work of A. von Thunen.[70] He drew up a conspectus of the land uses around an isolated town on a flat plain and pointed out the rings of intensity of use that would then occur as the costs of production and transport varied with distance away from the market which the town represented. Obviously, market gardening for perishable crops would be near the outskirts of the town, whereas forestry might occupy an outer zone. Modifications of the model could be made for variations in terrain. Other locational analysis models addressed themselves to e.g. the distribution of service centres and town size, the location of certain heavy industries, and the morphology of major cities and the functional zones within them. In the rediscovery and renewed application of this area of geographical study, the work of P. Haggett has been very influential.[71]

The significance of these models of regular distribution for environmental matters is clear: if there are ineluctable laws governing land use around markets, for example, then the intensity of environmental impact will vary with them. There will be lots of pesticide and nitrogen in the runoff near the town from the intensive market gardening; the forests needed for recreation will always be more than walking distance from the town itself. Similarly, if the laws governing the location of major industrial plant are inescapable, then it is nonsense to try and protect a population of wild orchids that are growing at the intersection of the cost/return curves. This type of argument has been extended on to a world scale with the notion of centre–periphery models of intensity of development, with the free market economies of the West and Japan forming a core and the South the periphery. Again, environmental manipulation may vary with location in the system. The reaction to this kind of Geography is that it is 'inhuman' and ignores the lifeworld of the individual and of societies. There has grown up, therefore, a 'humanistic school' of Geography which draws upon phenomenological models of behaviour; this is treated in Chapter 4.

Alternative geography

For the radicals of the 1960s and 1970s, the positivist movement in geography was seen as paralleling natural science itself in becoming part of a technocracy. Indeed since geographers often sought to influence public policy, they could be accused of lining themselves up with the ruling classes. It was pointed out (at first to a largely uncomprehending audience who had thought of Geography as being value-free) that much of locational analysis was underlain by implicit politico-economic systems such as free-market economics. This type of analysis usually ends up discussing the question of power in human societies: geography and Geography can both be seen as discourses 'of strategies and tactics of power'.[72]

Having made that clear, the radical geographers moved on to suggest alternative frameworks for analysis and, often, for action as well. The flagship of their endeavour was the journal *Antipode*, first published in virtually *samizdat* form, and now respectably under the wing of Blackwell of Oxford. In general, geographers eschewed the Cornucopian/Catastrophist dichotomy,[73] since most have tended to be optimistic and development-oriented by nature and upbringing, and opted for Marxism as their radical vehicle. In looking at problems of environment and resources, then, their analysis favoured the view that there was plenty for all provided the right social and political structures were present, a situation which clearly was not to hand since the West was over-producing for example food whereas in the South there was chronic malnutrition and often acute famine. The political analysis of the neo-Malthusian ideas about population, environment and resources current in the late 1960s by D. Harvey is a good example.[74] He extrapolates the notion of over-population into the occurrence of 'repression at home and neo-colonial policies abroad' and further suggests that no examination of such problems is ideology-free: any such possibility is itself an ideological belief.

At an even broader level, Marxian analysis of the neo-Malthusian ideas of the limits to growth opts for their dismissal as an outgrowth of the repressive nature of capitalism.[75] 'Limits' is seen as an artificial concept, designed to secure the ruling classes in their hegemony, and since nature and humanity are both historically interwoven, no such artificial and retrogressive structure need be entertained.[76] That the environments of the formerly Communist Party-led nations (such as those of Eastern Europe and the USSR) are as badly polluted as capitalist ones is seen as evidence of a failure to interpret and apply Marx properly, rather than as a limitation of his ideas.

Bring back physical geography?

Geography has sometimes been seen as 'Human Ecology' in an attempt to make sure that the 'physical environment' is not ignored.[77] In this case, the notion of there being limits to the manipulation of biophysical systems before they flip into some other state of temporary equilibrium is re-introduced, with the attendant proviso that this other state might not be culturally desirable. At some stage, the actual nature of such later states needs discussion, if prediction can in fact be made that accurate. In the case of the energy/economy interface which produces, for example, the increased concentrations of greenhouse gases in the atmosphere, accurate modelling enables us to consider what these limits might be. In other words, the separation of human geography out from an integrated discipline has made it possible to ignore the idea of limits to the possible detriment of all concerned.[78] But there can be no naive acceptance of the judgements of science here. These findings are, as we have seen, always provisional and indeed science here as elsewhere is only one input into the construction of a model in which action is sanctioned. The notion then of physical geographers setting a series of boundaries and the human geographers arraying their work within those is not acceptable: a truly human geography which both receives and gives meaning to those data is required.[79]

ANTHROPOLOGY

This group of workers has taken very much to heart Alexander Pope's dictum that the proper study of mankind is man. Founded in the nineteenth century wake of Darwinism and perhaps also in the desire to show that the theory of evolution can be applied to human culture as well as to species (and indeed to clarify the relationships of 'civilisation' and 'savagery') it has always focused in the end on the human community itself. That is not to say that environmental factors have been ignored, but for a large number of anthropologists they were for many decades subordinate to factors such as kinship, ritual and material culture. The type of anthropology which has contributed most to our theme is usually called ecological anthropology or cultural ecology and has been much affected by functional approaches to ecology such as systems analysis and energy flow modelling. It fits into the first category of a twofold division of anthropology today in which the first type of work is concerned with environment as niche and in which cultural strategies are adapted to increase the chances of survival. The second category sees landscape as subjectively perceived, as the setting for the symbols of culture.[80]

Cultural ecology

In his account of the development of this subject, B. S. Orlove draws attention to a number of stages that can be distinguished.[81] The first two he associates with the work of J. Steward and of L. A. White. Steward was especially interested in the cultural regularities in (pre-industrial) space and time that might occur in similar environments, and L. A. White combined two over-arching models. The first was his view that energy levels were a determinant of cultural evolution, and the second the application of Marxism. Later workers added other elements: the ideas of General Systems Theory permeated parts of anthropology in the same way that they had permeated ecology, as a way of describing a dynamic system with many interlinkages. Combining with this, was the biological notion of the carrying capacity of an environment for people. Culture might then be seen as a way of adapting to an environment's limitations, and the unit to be treated was a population (as in a natural ecosystem) rather than the more traditional social order of social anthropologists. This latter approach was called *neofunctionalism* and like its biological equivalent focused on regularities in ecosystem-level processes. In particular, the ways in which human populations functioned within ecosystems were discussed, usually as an examination of the mechanisms which linked social structures and material culture to the environment.[82]

The emphasis on process led to such studies as the relation of demographic variables to agricultural productivity as in the work of Esther Boserup,[83] the formation of adaptive strategies in extracting environmental resources, (concepts like the niche were transferred from biology), and on the lessons of history as elaborated by Marxists. But the major change from the past, occurring mostly in the 1960s, was a focus on ways in which behavioural and external constraints influenced each other: environment was no longer simply a background which provided enough necessities for people to go on living.

Applications

These newer views, coming as they did in the 1960s, interacted with the then very high level of public concern over 'the environment'. Anthropologists pointed out that humans were not behaviourally homogenous in the way of many animal species, and that an understanding of the ecological required a knowledge of the social. For example, where environmental decline and degradation were perceived as a problem, then anthropologists might be able to help with the social matrix of its causes. Notably, they might be able to identify the economically rewarding and psychologically satisfying behaviours that led to environmental decline and contamination, and suggest less harmful substitutes for them.[84] Essentially, they brought forward the idea (not now regarded as novel) that the human use of nature is inextricably bound up with the human use of humans; the remedies for the destructive use of the environment must be found within the social system itself. The leads from this towards a high valuation of indigenous knowledge in non-industrial societies are obvious.[85]

In a wider framework, R. F. Ellen praises the ecological viewpoint for its capacity to bring together something of a fragmented subject. Although he disavows calling ecology a discipline (preferring to think of it as a discursive practice or a problematic), he nevertheless sees it as bringing into one frame such issues as history, system, evolution and cultural adaptation.[86] In this way, then, anthropology might be one of the seed-beds from which new models of the human–environment relationship (at almost any scale) might emerge, though as yet this does not seem to have occurred.[87] One obstacle needing negotiation is the realisation of the deficiencies of anthropologists' description and interpretation of human communities: they are prone to be the describer's descriptions, not those of the described. This problem is not unique to anthropology; indeed it brings the discipline into line with all those other social sciences which find a tension between the perceived need to objectify and the demands of the greater reality of description from 'within'.[88] But here as elsewhere, the phenomenological approach has difficulty in transcending the level of appearance and of individual action. Yet if the subject can really enlarge the possibility of intelligible discourse between people quite different from one another in interest, outlook, wealth and power but who are all 'netted together' (as Darwin said of humanity and nature), then it cannot be but part of the larger enterprise of the construction of interfaces between humans and environment.

PHYSICAL PLANNING

The process of physical planning is quite literally the construction of an environment since the planners determine where the elements of the landscape should go: where there may be houses, which valued landscapes are to be protected from change, and where the roads lead. Following from the last section, it can be seen as the practical application of which human geography is the theory. Physical planning by governments at various levels is a feature of most nations of the industrial world. Even those with free-market economies have swallowed some of their economic liberalism to impose such a system, and the former CPEs positively demanded it. It is attempted in many developing countries but quite often the infrastructure for successful implementation is absent and so the actual plans may never come to fruition, especially where they aim to prevent something happening; this is not unknown in the West either.[89]

Orthodox planning

In the West, the dominating assumption seems to be that of a liberal and empirical Benthamism in which the greatest good for the greatest number is sought, within a relatively short time horizon.[90] The main elements which underlie the systems are the expression of individual preferences between hypothetical choices, and the conferment upon some body of the power of callousness in decision-making. Aggregate levels of preference, for a whole community for example, can be obtained by using compensation to buy off the losers in any programme of change.

This attitude has in most places superseded an earlier phase of no planning, in which the market was allowed to determine land use. From time to time this idea is revived even for the provision of amenity and beauty, qualities which most planners regard as being outside the range of the market. Thus, the argument runs, if wild country or even wilderness is desired then some entrepreneur will buy up such land and charge an entrance fee for it. Such ideas are especially popular in periods when economic liberalism is in the ascendant. Its opposite is the declaration of public interest, when an individual or group is allowed to decide what is best for the community, e.g. that there should be free access to protected wild country. Such groups carry out a kind of intuitive cost-benefit analysis but are unlikely to publish anything but the final decision.

The actual mechanisms of the planning process are mostly secondary to the politico-economic philosophies adopted.[91] It is possible to contrast, for example, a process which lays down a blueprint at the outset and then implements it without much deviation from the plan, with its opposite in which certain ways of procedure are laid down ('process mode') but not designated outcomes, so that planning is always a rolling process of which the outcome is by definition never certain. A plan can be rationally worked out and be as comprehensive as its professional input will allow, have been subject to alteration by public participation and adopted for a definite period into the future; it can also come together only in a fortuitous way as several sub-departments grope their way forward a year or two at a time in an uncertain political situation. But these ways of proceeding are nearly always subject to the overriding considerations of the prevailing ideology of who is to benefit most from the planning process. It is in theory possible to be purely functionalist, with planning directed at the needs of various groups as they arise, but this is in turn antithetical to notions of comprehensive planning which are dear to the hearts of professionals, who exemplify thus the western world-view about control over the environment.

Heterodox planning

The paragraph above has enough key-words in it to indicate that planning is never very far detached from political ideas. In general, moderate left-wing ideologies espouse planning, of the landscape just as much as they like economic planning, whereas the right wing is convinced of the superiority of the free market as an allocator of resources such as land for housing, parks and transportation facilities. More basic Socialism, however, is generally contemptuous of the general run of planning mechanisms since they are seen as being captured by one or other ruling group in order to perpetuate their hegemony over economic life. In the West at present, for example, Trotskyists would see the middle classes as having power over the workers in terms of their control of the planning mechanism at both the elected representative level and that of the professional officer.

On the other wing, there is a kind of High Tory view of planning, especially of rural areas. This is perhaps much influenced by Rousseau in the sense that it thinks of access to rural and 'natural' areas as being a necessary condition for the general happiness and indeed spiritual well-being of the whole population. This type of attitude, which is especially tender towards relics of the past, is willing to impose such values upon the population for their own good and without regard to cost. It is not surprising that the *Reichsnaturgeschutz* law of 1935 identified 'nature in wood and field' as being part of the inheritance of the German people; here it formed part of the mystical nexus of culture, race and place that was integral to the ideology of National Socialism.

Superficially, then, it might be possible to view planning as a positivist science in which values and political theory had no place. There is an enormous body of planning theory and practice which is rather akin to the textbook and the laboratory of the experimental scientist. But the whole direction of the system is determined by processes which have their roots in political-economic theories about values ('In the end, planning is a matter of judgement based on a system of values'[92]) and indeed the nature of society, just as the experiments the scientist chooses to do, reflect a choice of values. And just as in other systems of this type the nature of the non-human parts of the system is likely to be at best imperfectly understood and more likely to be low in the hierarchy of consideration. 'Environmental planning' is often a one-way process in which the planned get little say (even though they have advocates of a kind within most systems even if only via pressure groups) about their own future. In the last analysis, it is nearly always a way of translating ideas about power over nature into action, with a variable amount of concern for the pieces being rearranged on the board.

Behaviourist studies in environment-related planning

These studies are based on the school of psychology which suggests that work can be done on phenomena of human behaviour that are unambiguously observable and measurable. Such studies are said to be objective and with refinement will eventually lead to general laws of behaviour which will be predictive. The areas of application of most interest in this present context are landscape evaluation and research on response to environmental hazards.

Landscape evaluation

The nature of people's preferences for types of visual stimuli from features of their surroundings such as landscapes, types of building, paintings and photographs and even other people is to some extent a private matter. But not all questions of aesthetics can be left there since corporate decisions have to be made which affect the visual qualities of the environment: personal preferences have to be translated into public policy. If, however, we start with perception of the visual scene and the emotions and moods it evokes, then some generalities of response can be marshalled, which could form the basis for a transition from the purely individual to the totally public.

So the degree of generality encourages the thought that scientific modes of investigation can be put to good use and will yield acceptable and replicable results in the public sphere. Practical applications have been sought in areas such as housing developments where the residents' perception of safety, for instance, can be affected by visual stimuli; in Environmental Impact Statements where not only the material components of the ecological systems can be taken into account but the aesthetic whole as well; in legal systems where zoning decisions have to be made which affect the visual qualities of the region; and perhaps most used of all, in recreational areas and in development control where it is essential to control the rate of change (if necessary holding it to zero) of the appearance of the landscape. This latter activity, made necessary by the physical planning laws in many nations, has led to the development of an academic-planning discipline field called **landscape evaluation.**

This process depends upon agreement in a number of areas often reckoned to be subject to wide interpersonal variation, so it is encouraging to find overlap of the kinds discussed immediately above; not that such work has prevented the development of a multiplicity of approaches to landscape evaluation.[93] The main classes of model to date have been ecological, aesthetic, psychophysical and phenomenological. None of these models has any innate superiority over the others, though it seems as if the ecological one is really only suitable for largely wild country, and many would argue that the cultural decision to downgrade all forms of human activity is unacceptable. Similarly, the aesthetic formality seems to be so much part of the system of learned responses that it can be manipulated at will: go to any art gallery with a variety of periods represented and listen to visitors discussing the exhibits. For good or ill, however, the 'judgement' of the planner or architect or other professional (modified by contingent political pressures) is still the most influential in these matters of evaluation.[94]

But judgement need not always be only a matter of personal preference and found only in the lifeworld of the individual. Appleton has given us an example of a theory of landscape evaluation which is, he thinks, transpersonal and thus amenable to the kind of intersubjective verifiability that characterises science.[95] He suggests that the possession of pleasurable sensations in the experience of landscape are related to the existence of conditions favourable for biological survival. This 'habitat theory' is thus linked intimately with the Darwinian notion of evolution.[96] It leads on to his main hypothesis, which is called 'prospect-refuge theory'. It is based on the simple principle that an observer responds to a view in terms of whether he or she has an opportunity to see it (the notion of *prospect*) and to hide in it (find *refuge*) and in particular whether he or she can see without being seen. The capacity to achieve this fundamentally important stance, inherited from animal and pre-agricultural ancestors, brings about aesthetic satisfaction in us. The value of a landscape is apprehended aesthetically even when it ceases to be of strategic importance in survival.

Environmental hazards research

One of the fields in which the whole behaviourist approach to human evaluation of the environment is conspicuous is that of research into the cognition of and adjustment to environmental hazards.[97] These phenomena may be divided into **natural hazards**, which are events such as floods, droughts, storms, earthquakes and other phenomena over which humans have no control, and **environmental risks**, which are hazardous circumstances brought about by human activity, like atmospheric pollution and radioactive waste accumulation. One of the major spurs to the work has been the finding that while modern societies are increasingly able to manipulate their environments, they are becoming more than ever susceptible to the impact of natural disasters as well as the unwanted backwash of technological developments.

The founding principle of the behaviour-based work has been the concept of bounded rationality developed by H. Simon.[98] This argues that an individual constructs a simplified model of the world, in which he/she becomes a 'satisficer', looking for a solution which is good enough for the time being. We must expect, therefore, that differences between fully rational decision-making and actual adjustment will occur and in these interstices there will be damage to property and persons. A number of patterns of behaviour emerge from research conducted with this model as a foundation. The first is characteristic of e.g. an earthquake zone or an area of infrequent cyclones. The majority of people deny the risks involved or the significance of any possible damage and often assert that the hazard is not likely to recur in the near future. In the event of it actually happening, then the outcome is determined by the capacity of their livelihood to absorb the physical and human damage incurred. A second pattern is common to e.g. flood plain dwellers, drought-susceptible farmers or those living in the paths of lava flows. They are aware of the probability of recurrence and regard the effects as significant for them and their neighbours and relatives, but nevertheless decide that there is little they can do to reduce the impact of the hazard and so they either evacuate or seek outside help when the disaster strikes. Yet a third pattern is perhaps the more common. It too is found in flood and drought areas, and also in snow and wind hazard zones. Here there is some adjustment to the danger, by responding to emergency warnings, by anticipating losses and taking preventive action. This latter is seen in the case of coastal erosion, where expensive engineering works may be commissioned. An intensification of this kind of response may be seen as a fourth category in which individuals move away completely from the affected areas. This has happened in some drought-affected areas of Brazil and Australia.

National policies in the field of natural hazard management are relatively widespread. They span a range from the provision of after-the-event disaster relief, through event control which addresses itself to a particular hazard in time and space (like hurricanes or regular floods), to comprehensive damage reduction through, for example, national water use policies designed to cope with drought (as in Israel) or with industrial emissions which interact with climate, as in the UK clean air legislation of the 1950s which basically addressed itself to sulphurous fumes during times of atmospheric inversions. As might be expected, richer nations generally have better and more enforceable systems of ongoing government action than the countries of the South, though here the intermediate levels of warning systems, for example, may be highly effective in their provision of information.

At the international level, the first piece of behaviour of note is the existence of rapid and usually generous relief programmes following either a natural disaster (such as the famines following the failure of the rains in the Sahel zone of Africa in the 1980s) or a massive incursion of risk as with the Bhopal chemical plant explosion in December 1984. Such acts may be palliative both to those involved and to the consciences of the donors but help little in prevention. Here, a number of actions are worth considering at the global level, such as global monitoring and warning systems (e.g. for unusually heavy rainfall or likely drought), and data on the comparative effects and costs of different hazards might usefully be collected as a basis for the establishment of priorities. The training of personnel is another obvious area for international co-operation.

The work on the constructions of environmental behaviour under conditions of stress can be summarised as demonstrating in a cross-cultural fashion that certain abstractions can be made about human actions. These categories are *absorption, acceptance, reduction* and *change*, and they are brought about when various thresholds of *awareness, action* and *intolerance* are crossed. The factors which determine the mix of adjustments include the nature and probability of extreme events, recent experience, access to alternative resources and material wealth. The outcome of all this behaviour seems to be a trend in which the global total of damage to property will rise but deaths will fall in number. This differentiation will pick out a continuing emphasis on the loss of life in the developing countries versus damage to material wealth in the industrialised world. Those societies undergoing rapid development, especially in coastal areas, are likely to be hard hit on the property front as well. The overall outlook seems to be one of increasing loss from hazard and risk in the near future, though with the promise of a safer world in the future if the risks and hazards are not only better assessed scientifically, but societies evolve better mechanisms for adjusting their behaviour in the face of these threats to life, limb and property.

The above kind of systematisation is characteristic of a science-based approach. We should be wary, however, of supposing that the whole story has been laid out for us. In a radical perspective on hazard studies, Hewitt[99] argues that in fact most natural disasters are characteristic of the places in which they occur, that the risks they pose are part of 'ordinary life' in their locations and that they are known about rather better than many of today's social changes. In other words, much of the work is conditioned by a technocratic world-view characteristic of the centres of hazard research: the calamity is seen as a breakdown of the ordering of space. The alternative is to understand how prediction and containment of disaster may fail to deal with the real social problems. The notion of understanding, rather than prediction and control, comes from an 'inside-out' type of social science, which is discussed in the next chapter.

POSITIVIST SOCIAL SCIENCES: AN OVERVIEW

As with the natural sciences, we ask the question of whether we can read off a prescription for human behaviour from these investigations. Even more so than the natural sciences, the answer seems to be 'yes' since it is often the theory and practice of e.g. politics and sociology that lead more or less directly to the formulation and enforcement of law; economics is perhaps less directly implicated. The formation of environmental law is thus given a section to itself in Chapter 5.

If we try to summarise the relevance of the 'objective' social sciences for environmental relations, the first findings are basically economic. There is an enduring linkage between the planet's biophysical systems and the sustenance base of human economies of all kinds. These relations focus upon such features as population growth, technological change, the expansion of consumption and the expansion of production. They are all intertwined and difficult to separate, though the mainspring for change in the whole system is seen by neo-Malthusians as being population growth. Others, though, see the imperatives of production as being of greater significance since that imposes the 'need' to consume in the rich and provokes expectations of it among the poor. Unless the treadmill of production is dealt with, runs this argument,[100] none of the others will adjust in any non-degradational way.

The results of production and its concomitants are not, of course, equally distributed: there are all kinds of differentials in the gains from past environmental exploitation and control over future trajectories of that exploitation. Likewise, different groups run differential risks from the instability of ecosystems subjected to stress of both natural and human origin.

What commentary and what proposals for action might the 'objective' social sciences insert into these circumstances? What they have in common, as we can infer from the material which forms the bulk of this chapter, is that they tend towards being (or wanting to be) part of a policy science. They are happy to be part of a prescriptive process which leads to action and to the formulation of that part of civil law which forms a context for such actions.[101] They do this because they are judged in terms of their ability to provide rational explanations of processes and phenomena, much in the manner of the natural sciences. But if we follow one line of close examination of the ways in which natural science works, we see that explanation is not regarded as complete and as satisfactory unless it functions also as a prediction. And it is prediction which provides the foundation for control. Applied to the social sciences, it turns then into a kind of social engineering.

The argument proceeds further in the sense that the possibility of technological control (over environment and resources among other things[102]) is not simply contingent to science but part of the framework which constitutes the very possibility of scientific activity. Control is not, then, an option proceeding from scientific knowledge (and by extension that of the positivist social sciences as well) but part of the *a priori* assumptions which constitute the very possibility of enterprises themselves. So positivist social science is about control (in, for instance, guaranteeing production, influencing population growth rates, access to resources) so it has relevance and influence in modern society.[103] In the sense that most modern social sciences are products of the nineteenth century, there exists the suspicion that they, technology and industrialism all mutually reinforce one another.

Yet another interpretation of the relations between science, technology and the social sciences is that basic science provides knowledge from a set of sealed-off, internal, processes. This is applied as technology and this in turn is a forcing function for social and economic change. This belies the nature of the natural sciences as products of a world that is not neutral but which is guided by social, economic and political objectives. As H. Newby has put it, 'Science changes society, but it is also socially constructed'.[104] This presumably means that the consideration of environmental concerns is at least as much a moral discourse as it is a matter for the natural sciences.

So there seems to be scope to add at least one different kind of social science:[105] one which needs to account for the felt needs and for the sufferings of members of various social groups or even indeed of individual humans. This has to be grounded in the self-understanding of the 'actors' and hence must include participation by the actors themselves. What is produced, therefore, is not a picture of a social order and its workings but a catalytic agent of change within the complex it analyses. To do this, knowledge of a different kind is needed: that which comes from within i.e. that which is reflexive and interpretive and which deals with meanings rather more than with numerical magnitudes. In that task, though, the insights of the subject matter of the humanities (especially the visual and literary arts) have to be added to even a different kind of social science.

4

THE LIFEWORLD

Objective social science has critics. They argue that classification, systemisa-tion and theory-formation are inappropriate to the richness and diversity of the human condition. In the case of environment, the individuals' construc-tion of it, vertically as they grow up and horizontally as they relate to wider groups, can easily be lost. So the idea of the 'lifeworld', centred on a self, provides an alternative. This has been the subject of a great deal of philosophical discourse as well as empirical research, and is given further variety by the work of creative artists, some of whom treat environment, nature and place very seriously.

In this type of work, we start with the way in which an individual makes an environmental construction through the processes of perception and cognition. Perception is the term given to the neurophysiological process of the reception of stimuli from a person's surroundings.[1] In this process, sight is generally thought to be the major element, but other senses such as hearing and smell may also play their part. Recent research on the alleged health effects of living close to high voltage power lines has raised the possibility of electromagnetic waves also being an environmental stimulus. Perception is usually regarded as being immediate, i.e. it follows directly upon the stimulus, and is stimulus-dependent since the nature and very presence of the perception depends on the existence and type of the stimulus.

Cognition is the wider personal context of perception.[2] It is not necessarily immediate in the same way, since it constitutes the means of awareness that intervenes between past and present stimuli and the behavioural responses of the present and the future. It cannot therefore be easily disentangled from perception although the latter is sometimes regarded as a subset of cognition. The whole complex of cultural response such as memory, experience, values, evaluation and judgement are present in the processes of cognition with the result being a construction of environment which is perhaps analogous to a map of a landscape: a representation but not the terrain itself.[3] Thus environmental cognition can refer to elements in the environment, to events, to patterns and concepts, and to elusive qualities such as sentiment, personal meaning and collective symbolism.

Put together, perception and cognition are sometimes referred to as environmental 'knowing', though the use of the term cognition is probably more common. It is clearly a dynamic process which receives and organises the environmental information which helps individuals through their daily lives. Cognition is likely to vary from individual to individual and thence from group to group though most such units seem to have enough in common between their cognitions to make it possible to co-ordinate thought and action. That environmental cognition has implications for public policy is obvious. There is likely to be a discrepancy between words and deeds, for example. Methods of gauging public opinion (like polls) may founder upon the range of cognitions present, especially when hypothetical choices are presented. And like all cultural features, a ruling cognition can become the foundation for apparently rational decision-making, eclipsing all other outcomes.

Cognition and perception lead to behaviour itself, which we may regard as the taking of action in regard to some environmental feature. The range of behaviours is very wide; some are very localised and individual such as the choice of scenery on a walk, whereas others affect life much more basically as with the recognition and avoidance of earthquakes when locating built structures. Tied in with all the other influences there must be, too, an input from cognitive sources which contribute to the formation of major world-views like those of the twentieth century western free-market economies.

REFLEXIVE STUDIES OF BEHAVIOUR

This section acknowledges the centrality of the human individual in any studies of behaviour. Although it is in the scientific tradition to iron these out in the measurement of behavioural outcomes, it remains that there are psychological differences between individuals (leading, for example, to differences in physical sensation when exposed to the same stimuli), that there are differences in perception and cognition which are related to age, and that there are gender differences.

This leads to broader themes such as the relation between cognition and the establishment of a particular culture; the different ways in which a place appears to a visitor and to a native; and indeed to the whole question of attachment to place, which Yi Fu Tuan has called *topophilia*.[4] The concept of attachment perhaps reaches its extreme form in tourism, even to the point where the place cannot be very much like the prior cognition of it, when the travel company's brochures for instance are the source of the initial perceptions of that charming and unspoilt fishing village not all *that* far from Benidorm.[5] We ought to note in passing that although the individual's cognition of place is central to any construction of environment that he or she may make, the perception of time is also likely to be important. This may form a general and scientifically analysable constraint within a person's life[6] but it may also have a deeper individual meaning which transcends its objective chronological reality; experienced time, for example, may provide a different perspective. At the shallowest level of experience, we all know how 'the time just flew by' on some enjoyable occasion, whereas the wait for the next bus can seem interminable.[7] And individuals vary greatly in their capacity for *carpe diem*: some people seem to live only in their Filofaxed future whereas others seem to stretch out every present moment as if there were not likely to be many more.[8]

The self and the environment

The main school of thought that has attempted to get to grips with these relationships is that of **phenomenology**. This begins with the inspection of an individual's mental processes but all assumptions about the causes, consequences and significance of these processes are eliminated from the enquiry. What is being sought is 'essence', which is the defining properties of a thing or a process, and thereafter the meaning to the individual. It follows therefore that two people may detect different meanings of the same essence and their sense-data are held to be the significant facts, not the object or process itself nor the descriptions and categorisations of it that follow the rules of natural science. Out of such an obvious but possibly elusive set of concepts, different types of phenomenology have been demarcated[9] and in our context the most interesting seem to be:

- Descriptive phenomenology. This is the way in which the phenomena of the external world are presented in the lifeworld of individuals.[10]
- Hermeneutic phenomenology. This deals with the interpretation of meanings which are revealed as part of the conscious and unconscious lifeworld of an individual.

True phenomenology is therefore first-hand; it is my experience of myself and yours of yourself; all other interpretations are by definition second-hand. Communication may well be a problem because of the ambiguity and imprecision of language, which is still the only way in which highly subjective matters can be communicated to another and which can, as we all know, be misinterpreted without difficulty.[11]

In the context of our present concerns, phenomenology applies itself to such questions as 'how do environmental elements give and produce meaning in human lives?', 'how do humans organise their lives spatially in response to what stimuli?', 'how does a rural area contribute a different personal experience from a city?', 'are the five senses separate or synasthesic; is the visual sense as dominant over the others as our writings about place and environment suggest?'. Perhaps its most important assertion is that any person–environment relationship is illusory: people are immersed in a net of stimuli and response, including those of the body as well as the mind.[12] Though in many ways highly attractive because of its closeness to what humans apparently experience, phenomenology has faced severe criticism and not only from natural scientists. The findings, for example, are difficult to evaluate because no criteria for evaluation can exist (sample size is always one, and mathematics is often no better than ordinary language for communicating outcomes). Like much other work in the humanities, the results may not be cumulative and the field lends itself as much to preaching as to practice.

Attempts to spread these notions beyond the merely personal are particularly associated with the intellectual movement known as **hermeneutics** and its leading advocate, the German social theorist Jürgen Habermas (1929–).[13] Like the phenomenologists, Habermas gives up the illusion that humans can climb to some vantage point (like science or philosophy) that is somehow outside society, culture, history and the individual. There is for him no external reference point which guarantees objectivity or which provides a basis for description which has no presuppositions. But this is not a recipe for relativism, in which no one action is better than another. The hermeneutics consider that the individual, the community and the culture can achieve a 'fusion of horizons' in which there is a continuity of meaning beween tradition and the interpreter. Meaning and validity hence arise together in the process of interpretation. What is needed is a transparency of all the participants in which communication is free of domination by any one party and all rational subjects are involved in determining the shape of their lives. In all of this, as for many phenomenologists, language is central since it is the medium through which we exist; reality happens in language. Hermeneutics thus undertakes the difficult task of seeking an understanding (*verstehen*) of mentally meaningful subject matter, which includes our various environments. But it is very difficult to translate any of the findings into policy and collective action since in general it is just the individual who understands the world better as a result of looking reflexively at his or her own lifeworld. As a methodology, it is often accused of ignoring the material imperatives and constraints produced by others and by the hard facts of existence and thus some come to see it primarily as a critique of positivism in the sense that it infuses some of the feelings and the reality back into studies which by their abstractive and generalising nature have lost touch with what life is really like.

The intellectual tradition of most value in this context is that elaborated by Edmund Husserl (1859–1938) and Maurice Merleau-Ponty (1908–61), and which is applied to the present problems of the Catastrophist outlook by N. Everden.[14] All of these gentlemen wish to look beyond but not subsume the individual, and the starting-point is a naivety of mind which precedes 'knowledge' (in the sense of full cognition) in the way that the countryside precedes Geography or that wild cherry blossoms precede

Botany. This experiental knowing must precede the kind of observation that is informed by science and subsequent analysis; the cause of things is not at issue (nor perhaps even a subject for comment) but the meaning is. So the first requirement is not explanation but description, which brings with it the formidable challenge of describing the world of another person and then seeing if those experiences have any more general meaning for humankind, i.e. do they provide an inclusive overview of the experience of being human?

To carry that question further was one of the contributions of the philosopher Martin Heidegger (1889–1976). If we consider the statement, 'the sky is blue', then there is the remarkable verb the sky *is*, which contains the implication that the sky might not be blue. It is a condition of being human that we are aware of the possibility of non-existence, whether it is our own mortality we consider, or the extinction of a species, or simply some thing which has not come into our experience. Descartes, in his famous phrase, 'I think, therefore I am', never elucidated what 'I am' actually meant. So to understand Heidegger is to enter into an alternative cosmos of being and meaning. There is no place, for example, for terms for things, only essences: we have to imagine that the visible presence of a tree or a person is removed and only its actions and relationships remain. Mankind becomes a state of Being[15] which becomes visible only in its relationships with the rest of the planet through its concern for all the other Beings on it and of it. (The language used is difficult: 'Being' is a translation of Heidegger's term *Dasein*, which can also be rendered as 'being there'.[16]) The world then becomes not a set of objects but of significances and meanings, among which is the possibility of non-existence. The extinction of a species or a habitat then is to hear one's own name called into a state of non-being.[17] In the end, Descartes' motto becomes transformed into 'I care, therefore I am': since we inhabit all those parts of the world that are open to understanding (which comes via experience) then we have the ethical duty to be the Shepherd and Custodian of that world, maintaining a respect for all the other lifeworlds that it contains. One result of this world-view is that space becomes a qualitative mosaic (and not a rigid and confining quantitative geometry as the positivists might construct it) which we know as individuals and as a species. More than most, if not all, living things we have a freedom of 'world-formation' and thus can modify our behaviour in a way not open to other species. We do not therefore *have* to encounter the world simply as a field for the use of tools: a feeling of uncertainty and wonder is as valid a reaction to the world as the completion of a new road. Whether this somewhat rarified concept can be translated into an actual environmental ethic is a subject for Chapter 5.

As with 'objective' views of environment, the reflexive approach has generated its mainstream and its alternative practitioners. The French philosopher Maurice Merleau-Ponty, mentioned above, is a fertile source for those wishing to establish a mind-based holism for the planet. He particularly sought to establish languages as a method of communication between Earth and her inhabitants: a clarified description of other embodied forms moving at rhythms different from our own. David Abram writes that 'this language we speak is the voice of the living Earth itself . . . the diversity of the various languages we speak may well correspond to the . . . diversity of Earth's biosphere'.[18] Proponents of the Gaia hypothesis transfer such an interpenetrative view to communication in the whole biosphere. Perception is an attribute of the whole biosphere (as a whole) rather than any species within it. These multiple forms of perceptual experience may collectively, it is argued, constitute a single global experience that has its own power of sensing and, perhaps most importantly, of creating.[19]

Giambattista Vico's *scienza nuova*

Older than the German tradition being discussed, but only recently being revived, is the humanist thought of G. Vico (1668–1744), who lived in Naples.[20] His contribution is also to stand Cartesian dualism on its head, by means of a two-fold distinction between *scienza* (science, which might lead us to truth) and *coscienza* (consciousness, which might lead us to certainty, a lower category). Vico argued that we can only engage in *scienza* when we know the causes of things, and further that we can only know these when we ourselves are the creators of what we are studying. Hence the world presents a continuum from the man-made, which is accessible to *scienza*, to the natural which is only knowable as *coscienza*. In environmental terms, it appears that we may know the truth about the city since we ourselves created it, but about pristine ecosystems we can only aspire to a consciousness which although dignified with the name of certainty is presumably always a transitional one. This lineament of change is emphasised by Vico's insistence on self-examination as part of the process of knowing: he thought that each of us went through a set of unfolding stages in our lives which affected what we thought, as indeed did the whole of human society.[21] To a great extent, therefore, Vico's thought bridges the gap between Cartesian dualism and the more personal aspects of reflexive philosophies. The centrality of the human is affirmed but the possibility of natural science is also admitted since it is needed for those aspects of the world we know only by *coscienza*. Just how to formulate the mix of the two for the study of environmental impact seems at present to be a task for future devotees of this unusual Neapolitan.

ARTS AND ENVIRONMENT

The cognitions so far discussed have mainly been those of response: taking what is and searching for meaning. But the individual is capable of a different activity, namely that of creating. In our present terms, he or she can put forward for consideration a new construction, in words, images, sounds or materials. This is the field of the arts.

At the outset, we have to acknowledge the breadth and fuzzy edges to meaning which beset any discussion of 'the arts'. To call them, as does one modern dictionary, 'imaginative, creative and non-scientific', does not get us very far. To think of them as being essentially aesthetic in appeal rather than practical in application excludes interfaces like architecture, and the whole question can be muddied by the role of initial design in the process of mass manufacture.[22] Yet there can be no doubt that many writers, creators of visual images, architects, gardeners and musicians have dealt with the environment, some creating cognitive constructions for themselves and an audience, others actually creating physical environments like the park at Versailles or those experimental wind-tunnels called shopping precincts. Most of them have dealt with the environment by calling it 'nature' or 'landscape', though for some it has been something of a static background and we shall perhaps want to look more closely at examples where the dynamic interaction of humans and nature is the subject of artistic expression.[23]

For present purposes, I shall adopt a simple but arbitrary scheme of what is to be taken under discussion,[24] thus:

1 Practical arts: those which *primarily* serve a useful function but which may well also appeal to, and perhaps try to mould, our aesthetic senses, like architecture or automobile design.
2 Fine arts: *primarily* designed to appeal to the senses and not to go beyond that in terms of utility. We should include the visual arts, the auditory arts, literary art, and presumably mixed arts like drama in all its forms which may combine spectacle, words and music.[25]

Few readers, I trust, will have difficulty in identifying arts which transcend these classificatory boundaries or perhaps fit into none of them: how about making love, for instance? Here, though, we shall concentrate on those arts which have contributed most to our constructions of environment in recent times and then discuss some of the others at rather shorter length.[26] There is clearly another book to be written which tries to put all these expressions of sensibility together for our present times.

The visual arts and the environment

The predominant art we shall discuss must necessarily be that of painting and drawing, but there is nowadays no case for omitting a consideration of photography; sculpture is also primarily visual in nature and there are aspects of mixed media like video, television and film which are appropriately mentioned under this heading.

Our earlier discussions drew a distinction between constructions of environment which were largely individualist (some types of phenomenological approach for instance) and those which were collective, in meaning the same to everybody, as with positivist science. To some extent, there is an analogy in the case of most arts, but especially the visual ones: there is first of all the personal vision of the artist which remains more or less just that. It forms a personal construction of his or her external world which does not connect outside at all. By contrast, the visions of some executants connect outside a great deal. They may influence other artists and so form a school, or at the very least a fashion: a major example would be the spread of the idea of perspective in painting and drawing during the Renaissance. Second, they may influence public ethics and possibly public policy: in the eighteenth century having your portrait painted in front of your own landscape park and then exhibiting the picture in your house was clearly setting the seal of approval on large-scale property ownership. The earthrise photographs of the 'blue planet' are said to have induced a new consciousness of the uniqueness and fragility of Earth. However, we must admit that there is no objective and quantifiable way of assessing the degree of outward connectivity and no doubt many artists themselves would argue that it would be meaningless if there were: even in science, citation indices are limited in value.

Painting and drawing

In most cases of these arts, the surface of the environment is presented to us in the form of landscape either with or without figures.[27] In the course of time, oil painting has been the major medium for these scenes, but pencil and ink drawings, watercolour, mixed media (e.g. pastel and ink) and collages are all employed.

After the seventeenth century in the West, the landscape alone is presented to us as worthy of depiction, whereas before then it was mostly a background to human activity. According to Kenneth Clark,[28] the tacit question in landscape painting is always, 'what is man's relation to nature?'. In answering this we can see that the landscape can be an objective record of its appearance at a particular time and place and so its meaning can indeed end there; it can also be a set of symbols which points us to the dynamics and the interaction of humans with their surroundings.[29] In other words it is seen as part of their experiental lifeworld and possibly also then of ours. These latter are for our purposes the most interesting, though they are not necessarily the most aesthetically appealing in received opinion or valuable as investments. Here I shall try to give a few examples of paintings which show us various ways of showing landscape and some of the reactions that have arisen. The selection will be highly partial in many ways and there can be no pretence of comprehensiveness.[30]

It seems generally agreed that in the West the Renaissance of Italy witnessed the transformation of the role of landscape in painting. In the earlier pictures a formal and not very realistic landscape is the static backdrop to the human figures and the context is mostly Christian. Some change is attributed to Leonardo da Vinci (1452–1519) whose middle-ground vegetation was clearly recognisable but whose mountains have the air of cosmic forces from another time. Giovanni Bellini (?1430–1516) and Piero di Cosimo (1461–1521) took a further step into realism (albeit tempered with a strong mythical element in the latter case), with trees, flowers, mountains and towns all forming a totally credible whole with the figures.[31] For many critics the key painting seems to be Giorgione's *The Tempest* of the late fourteenth century.[32] Like some of its predecessors, it is an inviting landscape: there is mostly discovery and delight to be had in it; it is rounded and soft in form and the dimension of fertility is emphasised by the figure of the woman suckling a baby. In the background, though, there is a storm and there are also classical ruins associated with the onlooking soldier. Interpretations of the scene are many and various but we might read into it a unity of human figures and their surroundings in which birth, fertility, destruction and change are all inextricably interwoven but presented as good: a kind of moral ecology, perhaps.

In the Netherlands, the land and sea came to form a dominant element in many pictures, subordinated only to the sky, of which there is a lot in many such flat terrains.[33] To some degree, this development was prefigured by Jan van Eyck (*c.* 1390–1441), whose *Madonna with Chancellor Rolin* is set in a landscape of such meticulous detail and calm order that we cannot be invited to think other than all is well in the cosmos and that everything is in its rightful place. A further statement of connectivity comes in the famous landscapes of Pieter Breugel the Elder (1525/30–1569), especially his *Months* cycle. The winter picture, *Hunters in the Snow*, is especially evocative: his mountains are reminiscent of the conventions of the early Italian Renaissance, but the rest is a stunningly humanised landscape, which fact cannot be disguised by even the snow cover. Here again, the very detail invites the viewer to find himself a place in it: none of the figures nor the other locations are overbearing and dominant to the point of dwarfing the onlooker as in many of the classic paintings of Italy.[34]

The Dutch and Flemish landscapes of the seventeenth century are the classics of depiction of environment.[35] Huge sweeps of sky, flat dark terrains with the gleams of many water-bodies, sharp relief only from features like sand-dunes, all convey a sense of the dominance of the environment which the homely cows and cosy cottages do not dispel. A deterministic interpretation might well focus on the precariousness of the Low Countries at that time as far as the flood hazard was concerned. Philips de Koninck (1619–88) provides an example in his *Landscape: a View in Holland* where a more or less total inundation of everything except the raised left foreground can readily be imagined. Humanity in these pictures is in awe of nature rather than at ease in it, perhaps. However, it is not difficult to find examples of dominance by the works of humanity in this type of painting: the classic example is probably Meindert Hobbema's *Avenue, Middleharnis* of 1689. Its linearity of roads, tree-planting and ditches all add up to an imposition of man-made order on what is, presumably, a landscape of reclaimed wetland. The linearity prefigures the emplacement of the railways 150 years later. Heroic posturing in the seventeenth century is typified by Claude Lorrain (1600–82) who worked in Rome and who formulated idealised landscapes with classical figures in a rather theatrical manner (with foregrounds like a proscenium) and who greatly influenced English writers like Pope and the landscape gardeners like William Kent and 'Capability' Brown. Their landscape parks were often the settings for eighteenth-century portraiture of the country gentry sitting firmly on their property (with the occasional intervention of a horse, itself a desirable possession of the newly come-up, in the manner of the Porsche of today's yuppie); the movie *The Draughtsman's Contract*, set in the seventeenth century, has much the same to say about landscape and tamed nature as property.

In the present context, much of the eighteenth century and the early nineteenth seems to be an age of realism in which all landscapes along with portraits strove for a mastery of all the presented detail, a function which was to be taken over by photography. The main variant came with artists like Constable (1776–1837) who were adept at showing the changing play of light across the land and its relations to the form and movement of clouds. Nevertheless this type of naturalism attracted the opprobrium of Horace Walpole, who referred to the 'drudging mimics of nature's most uncomely coarseness'.[36] Getting beyond this surface detail, and responding to the new relationships posited by the advent of industrialism, there is considerable interest for us in the work of J. W. M. Turner (1775–1851). Three of his pictures construct environmental relationships. The first is of a sailing whaler with her catch (*Hurrah for the Whaler 'Erebus'*, 1845) in which the old order predominates: though men are set against nature they are clearly having a hard fight to prevail;[37] the second is of the *Fighting Temeraire* being towed to the breaker's yard by a steam tug (1838) which can be seen as the end of the pre-industrial world and the triumph of the new force of steam power (Figure 4.1). The undisputed emergence of this force is in the famous *Rain, Steam and Speed* (1844) where the rigid lines of the railway bridge allow the train to forge unchecked through the weather at a then unheard-of velocity.

The way in which painting and drawing dealt with the industrial revolution, with all its consequences for environmental change, is most interesting. Nowhere are the relevant years more concentratedly covered than in the appropriate rooms, arranged chronologically, of the Musée d'Orsay in Paris. Here, threading our way through the younger schoolchildren

Figure 4.1 J. W. M. Turner, *Fighting Temeraire* (1838). The ship of the line is being towed to the breaker's yard by a steam tug: a 'defining moment' in the transition of a new relationship between humans and environment

Source: Reproduced by courtesy of the Trustees, The National Gallery, London.

Figure 4.2 Claude Monet, *Train dans la Campagne* (1870–1), in which the train is partly hidden by natural phenomena, though sending signals of its presence out to a wider audience.

sprawled on the floor and the American teenagers boasting that they have
been round without looking at a single picture, is the visual art of an
industrialising nation (the Musée covers the period 1848–1914) consisting
largely of the Impressionist and post-Impressionist painters. The key
observation we may make is that to them at least there was virtually no such
process: V. van Gogh portrayed factories at Clichy once and near the end
of the sequence a canvas of Cardiff Docks' railways at night (by Lionel
Walden) hangs well up the wall; Pissaro and Sisley brush occasionally
against chimneys and factories but overall, they clearly dismissed them from
their artistic endeavours. Gauguin, after all, went to Tahiti to get away.[38]

The exception seems to be Claude Monet (1840–1926), and this is only
partial for there seems to be a difference between his earlier work (up to
c. 1890) and the years thereafter at Giverny which produced the water lilies
and the *Ponts Japonaises*.[39] But between 1870 and 1890 there are several
canvases which record the impress of the new way of life. In some it is
peripheral in the sense that it is tucked away into the distance in the form
of a railway bridge (as in *Le Bassin d'Argenteuil*, 1872). The clearest
statement of this relationship is in *Train dans la Campagne* (1870–71), where
the fore and middle grounds are occupied by a grassy meadow and dense
woodland respectively, and across the rear horizon is the train, the locomotive being invisible and detectable only by its smoke plume (Figure 4.2).
In *Le Pont de Chemin à Fer à Argenteuil* (1874) and *Train dans la Neige*
(1875), the new technology is brought further forward. In the former (of
which there are two versions), the straight lines of the bridge pillars and
decking, viewed from below, dominate the picture, being laid ruler-like
across the more rounded outlines of the near river-bank and the tree
opposite. The second picture is mostly straight lines: the track with its train
is paralleled by a fence and a line of equally spaced and hence clearly
planted trees. Trains themselves are even more dominant in the sequence of
pictures of Paris railway termini, of which the Gare St Lazare group (12
altogether) of 1877 are typical. In one of them, the terminus roof encloses
the entire picture in an iron web; beneath it the trains and buildings seem
to be held down. However, the atmosphere is clean and light, with the
smoke from the locomotive a rather pleasant Wedgewood blue. This
contrasts with a lesser known picture of Monet's, *Le Ruisseau de Robec à
Rouen* (1872), which has an overall flat sludgy-brown colouring, including
the water of the *ruisseau*. There is a mélange of small industrial buildings
blocking off the perspective and a greenish tree is the only sign of life. In
the Gare St Lazare, we feel the terminus of Monet's involvement with
industrialisation, against which he reacted to the point of painting water
lilies and Japanese bridges, with even the latter eventually losing any
sharpness of outline and becoming merged with the vegetation around.[40]

Two post-Impressionists are of interest here. As mentioned briefly above,
Vincent van Gogh (1853–90) painted an industrial scene in his *Usines à*

Clichy in 1887. This contrasts and compares with, for example, his 1889 *Cornfield with Cypresses*. In the latter the swirling brushwork seems to link thematically the vegetation, the hills and the sky, rather in the way energy might flow through the biosphere and atmosphere. In the industrial scene, however, only the foreground field gets this treatment: beyond the horizontal fence the technique is far more conventional: the work of humans is completely divorced in space and technique from that of nature. In the case of Paul Cézanne (1839–1906), we see in his later works the use of planes of colour to bring the structure of the landscape forward to us where we can see (to some extent) how it might be made up: examples are *La Montaigne Ste Victoire* (1902–6) and *Mountains seen from L'Estaque* (1878–80). With the absence of human figures, we ourselves have to relate to the terrain: I find it full of latent hostility and highly unyielding. The later work of Graham Sutherland (1903–80) follows in this vein: it sought to show the inner structures of nature and the dynamics of change and was never afraid to be uncompromising.

Painting and its allied arts are not simply visual portrayals of a scene. They can get into the bones of their subjects and hint at or lead us towards their functioning, e.g. as economic property or as ecosystem dynamics. They can also be normative, i.e. telling us what *should* be. According to Cosgrove and Thornes,[41] art can encompass truth, beauty and relation. The latter gives us a sense of value or meaning. John Ruskin (1819–1900) thought that landscape should be subject to exegesis rather like a text and that art was one of the ways of accomplishing this. One result was, for instance, that beauty was not necessarily in the eye of the beholder but was intrinsic to a scene because that scene was symbolic of the divine purpose. All works of man were to be tested by their concurrence with, or subjection to, that end. Ruskin was highly aware that the Romantic world of Wordsworth was passing and that storm-clouds of an industrial nature, with links to pollution, were forming; he was also confused by the work of Turner and the Impressionists. In their wake and that of the industrial revolution, he sought to reformulate a harmony between people and nature. He was appalled by nineteenth-century capitalism and thought that a purely materialist basis for a human world was an impoverished thing. By contrast, though, from the High Renaissance of Venice onwards, pastoral landscapists have tried to transport their viewers to an idealised set of locations: fresh, green pleasances which convey a sense of harmony between humans and nature.[42] Nostalgia for a Golden Age and an Arcadia has existed side-by-side with the sterner moral invocations: art can try to reassure as well as disturb.

Sculpture

For reasons of space as well as relevance, discussion must be brief since sculpture is not one of the major contributors to environmental awareness in the arts. True, the subject matter may be 'nature' in one or another form, and true also that the artists may claim a close identification with natural materials such as stone or wood, or to be catalysed by a natural form such as a pebble or a sea-shell. It is also the case that some sculpture is often shown deliberately in an outdoor setting which places the piece in an environment, just as a real human would be placed: the work of Henry Moore (1898–1986) is often seen thus, and to great advantage. There seem, however, to be few sculptors who confront the question asked of the painters, 'what does this tell us about the relations of humans to nature?'. Moore is one of these, for he apparently looked for correspondences: for him there were universal shapes to which everybody was subconsciously conditioned if they did not deliberately cut themselves off. Natural forms excited him if they complied with the form which he was seeking at the time or if they were sufficiently similar to such a shape but yet provided an extension of it into a newer form. A more abstract but equally valid line of expression runs through, for example, Constantin Brancusi (1876–1957) and Barbara Hepworth (1903–75). Here there is a reduction of form to a simplicity which allows the sculpture to merge with its surroundings and thus posit a non-duality of it and its environment. (Brancusi, we may be reminded, was much influenced by Tibetan mysticism.[43]) This seems to happen even in a gallery with e.g. Brancusi's *The Bird* (1912) or *Bird in Space* (1919) but is formidably so in the case of Barbara Hepworth's works as set out in her own garden in Cornwall.[44] The lightness and etiolation of her figures (she was impressed by ballet dancers) and the frequent use of piercing to allow the background to become part of the sculpture, suggest a prefiguration of the mythic side of the Gaia concept or perhaps a rather austere and northern form of the cosmic dance of Shiva with which we are all familiar.

This leads us to twentieth-century abstract art, where the release from any necessity to depict surface form in a representational manner has allowed artists the freedom to explore and depict the inner structures of landscape, seascape or that of organisms.[45] Time has not yet winnowed out the work of say the post-1945 period in this respect though possibly the most memorable constructions are literally so: the actual landscape manipulations of Christo Jaracheff (1935–) like his *Running Fence* across much of coastal Mendocino county in California, or his swathing of coastal rocks in Australia in 1969. To a Green-minded world these seem like Promethean efforts to bring into the human fold the otherwise untamed world of nature and to put on it the kind of impress that we associate with industrialisation. The opposite tendency is that of Richard Long (1945–) who makes simple geometrical shapes (e.g. lines, crosses, circles) of the local materials, treading lightly on the earth, as it were, for he always walks to the sites.[46] These are then photographed for exhibition. The environmental relationship seems to be shown by the lowness of the sculptures, and the fact that they are transient features: like life, temporary aggregations of complexity on the route to entropy.

Photography

Like painting, photography is much more selective of the visual field than is the human eye, round which there is, for instance, no frame. What is put in front of us is therefore a partial view of reality but in the West today, any deficiency in that direction receives a counterweight from the sheer quantity of photographic images that we encounter. Given the advances in technology (e.g. autofocus) and the popularity of photography as a hobby, it seems as if there will be even more of them. For our purposes, though, we will omit the family album and slide-show[47] and encounter two separable types of the medium: 'art' and 'commercial' photography.

The posed portraits of the Victorian studio are often repeated in the outdoors. Partly this was due to technical limitations but also seems to result from a reinforcement of attitudes. The work of Henry Peach Robinson, for example, places girls and women in gardens or in gentle countryside, as evidence of that part of their outlook in which women were gentle creatures essentially decorative in function, and put nature in the background simply as stuff to be used for whatever purpose was in hand. The Yorkshireman Frank Meadows Sutcliffe, on the other hand, in his working class work from the coast did not romanticise what was a very hard way of life, even though there is little overt environmental interest in his pictures.

Among the most famous modern photography of landscape is that of the United States' citizen Ansel Adams (1902–1984), who is especially noted for his images of the Sierra Nevada (including Yosemite Valley) of California.[48] In these very strong black and white pictures the human presence is barely acknowledged and the forms allow at least two opposite interpretations: to some, there seems to be a distinct separation of elements due to the high resolution of the different tones; to others, the use of black and white causes a thread of continuity. Clouds, rocks and water all seem manifestations of the same elements. Whatever the message of his work, overt or subliminal, Adams' imagery has been very influential in the wilderness movement in the USA, being used often as part of its publicity material, to gain membership of conservation organisations or in waging anti-development campaigns. This brings us explicitly to a kind of 'environmentally conscious' type of photography. Now, most facets of nature are thus recorded, using miraculous equipment and highly developed colour materials: whole continents from satellites through to the micro-organisms in a drop of water and beyond are all fixed for all time (or as long as the film stock lasts) in numerous archives, both private and commercial. Mixed media are popular, too, using pictures along with writing or *objets trouvées* to make collages dedicated to particular environments. The recent work of

Figure 4.3 Fay Godwin, Path and Reservoir above Lumbutts, West Yorkshire (1977)
Source: Reproduced by courtesy of Fay Godwin/Network.

UK photographer Fay Godwin (see Figure 4.3) in conjunction with the poet Ted Hughes in a volume devoted to the South Yorkshire region of Elmet (a part of the southern Pennines which had a now declining textile industry) is instructive.[49] For her, the works of man (*sensu stricto* here, I think) seem to grow organically out of the largely gritstone landscape and are just as easily reabsorbed into it. There is a basic bleakness of world-view which echoes one of Hughes' poems:

> Heather is listening
> Past hikers, gunshots, picnickers
> For the star-drift
> Of the returning ice.

Here, the retreat from the industrialisation of the nineteenth century is seen as a microcosm of the whole progress of the post-glacial period.

In the commercial world, there is a plethora of images, together with the moving ones of TV commercials and the like. Here they are usually in a mixed form, since the picture itself is rarely deemed to tell enough of the story for e.g. advertising purposes.[50] It is a very versatile medium and can be manipulated to sell whatever image of the buyer it is desired to project. In doing so, it may also inadvertently convey an environmental construction as well: consider the way in which a benign nature seems to be a desirable adjunct to almost everything from shampoo to cars and wholemeal bread with no additives. The benign nature presumably reflects a tamed environment which is a supplier of materials: the whole western *Weltanschauung* in two minutes between programmes. When 'alternative' organisations use photography then it is usually much more direct and more polarised: animals, for example, are cuddly like pandas or in danger like fur seals or laboratory animals. If we doubt the power of the visible image, though, just imagine going through a day now where only words came at us from the media, or count the number of times in advertising where the information is validated because it appears on a computer screen: on TV there is a screen within a screen. But that is perhaps another story.

The literary arts

Literary mediation with our environment spans a great range of material, written for many different occasions, with the commonality that the writer believed in words as words, that they had power or some other quality worth putting into a public arena. So from the earliest writings that have survived down to this morning's newspaper, some words appear which have other than a purely utilitarian function. In terms of approach, there are probably more words written now about the science of the environment than from a literary angle; in terms of subject it is likely that love still excites people to the most verbiage, but prose and poetry span all of time, place and culture when they plug in to the play of forces that constitute the non-human lifeworld.[51] Since the dissemination of printing, most of the writing of the past is available for us, too, in as unaltered a form as it was originally, more or less; this is something we cannot say of the biophysical systems themselves, which have to be reconstructed by laborious means. We must therefore be selective in our approach and there will be something of a concentration on the nineteenth and twentieth centuries and on the English language, though not to the exclusion of all other times and tongues.

We have to go beyond simply dealing with 'nature' or 'landscape' as subject-matter, as we have said already. We have to transpose Kenneth Clark's question about painting and ask what, if anything, this piece of writing tells us about the relations of humans to nature. It follows that there must be an engagement between the two resulting in a genuine construction rather than simply the erection of a piece of static scenery before which the play is enacted. The difference is hard to tease out in the abstract but for the moment perhaps we can accept that we know the real thing when we see (or hear) it.[52]

The professionals have ventured into this field. The literary critics have become interested in landscape as, for example, metaphor,[53] the geographers in literature as a source of meaning for 'places',[54] to name two instances. The novelist Margaret Drabble has managed to classify, for Britain at any rate, the types of place in which, she says, 'every writer's work is a record both of himself and of the age in which he [sic] lives, as well as of the particular places he describes'. For her, then, there are Sacred Places, The Pastoral Vision, Landscape as Art, The Romantics, The Industrial Scene and the Golden Age.[55]

The prose of place

As mentioned above, the invention and dissemination of printing have meant that literature is rather different from the objects to which it refers, for it can be preserved in pristine form. The ravages of time which have altered Bede's rivers so that they no longer abound in fish have made Bede's words available in any bookshop. But the meaning of the words is subject to change: the climate of thought[56] will alter through time, and each individual reader will bring a new set of lenses to a reading, not the least formative of which will be all the other things that he or she has read. So the total written past is available for incorporation into everyman's worldview. Not that it is necessarily critical: does Proust's recollection of his childhood walks influence the siting of a new motorway in rural France? We may have leave to doubt it: there is little place for search for the past in a Cost-Benefit Analysis.[57] Even so, literature can give us powerful images: Coleridge's albatross and *Catch-22* have passed into the language.

The tension between the individual and the collective is nowhere more apparent than in this field; to what extent is literature a private contract between writer and reader, or is there some wider purpose to all those words? If only the former, does it leave us open to the view that this is largely the province of a conventionally educated élite for whom the experience of visiting Dorset is much enhanced by their knowledge of Thomas Hardy? Even more, if we confine ourselves largely to literature in this élite sense, are we open to the charge that the literature that e.g. promotes high technology (like much science fiction for instance) is overlooked in favour of a more backward-looking set of world-views, as in Margaret Drabble's list? Probably so, but to do the job thoroughly is an enterprise of greater scope than this work.

Bearing in mind some of these caveats, we can approach the environment in prose literature in a number of ways. We can classify its occurrences perhaps:

- where it is simply as background, as in the quotation from Auden on p. 187;
- where the subject-matter is a facet of the environment, like trees, the sea, woods and animals; and
- the commonest, where the environment presents itself as a *place* with meaning. It is not simply a set of objects to be repositioned by a planner, but somewhere that has experiential meaning for writers who have the inspiration and the craft to communicate this to others, as it were to get inside the landscape and take us with them.

The notion of meaning often invokes the role of history and for us this must include any relation of a contemporary story to those that have gone before and in particular those which have the status of myths since they are likely to reveal long-standing and cross-cultural truths about ourselves and our relations to nature. If we take as archetypal the Greek formulations (of what must already have been ancient mythologies) then some seem to have a direct environmental applicability. The legend of Odysseus, for example, seems to point to the idea that at the end of all exploration, the highest value is to come home and die quietly in our beds, having put to flight all those who were plundering the place while we were away in the hope that the owner (Owner, perhaps) would not come back. The uninviting story of Prometheus is often invoked by Cornucopians and Catastrophists alike, but for different groups there are altered emphases on the part that deals with paying the price of the new discovery.

Places change and so there is a constantly renewed market for a 'fresh look' at familiar and unfamiliar places alike, through the lenses of someone else's prejudices. This format translates very readily to television and so there is never any shortage (at least where there are publicly financed channels) of 'person-alities' wending their way through Victorian suburbs or the remains of medieval markets or the whole of the former USSR in six 30-minute programmes.

But it is in the field of imaginative writing (for which the term 'fiction' seems somehow inadequate) that we can be brought more abruptly into focus against nature or against place, since the everyday contingencies can be stretched and moulded at the author's command. Our tendency here is to think only of the 'serious' novel and short story and to disregard the fact that many readers get their wordage from other channels. This is especially true of children, of course, for whom a specialist literature exists and one which is by no means poor in its use of the environment. At one level, naturally, life is full of humanised animals (less often plants) of the Beatrix Potter type but there exist some very much more searching stories. Even A. A. Milne's characters exist in a variegated environment of the type that affords a kind of mini-paradise for the very young; its urban equivalent is the derelict area of former mine or factory. In World War II the bomb site was an immense adventure playground: see the movie *Hope and Glory* for an accurate (I can testify personally to this) evocation. Animals, though, can carry a much greater weight of seriousness than that, as is evident from Richard Adam's books like *Watership Down* and *The Plague Dogs*. Here, we are invited to confront the apparent misuse of rural environments and of laboratory animals head-on, without sentimentality creeping in to the 'languaging' of the animals. Modern ecology is used as the basis for David Bellamy's *The Mouse Book* to the point where ecosystem processes themselves virtually constitute the story and indeed the culmination is the death by predation of the fieldmouse, knowing that such an event is the proper way to die.

For a construction of a total imaginary world (though not totally imaginary), we turn to the famous trilogy of J. R. R. Tolkien, *The Lord of the Rings*. This was popular with a much wider group of people in the early 1970s, and especially with the College-age cohort holding environmentalist views. Its success is aided by the employment of a series of environments that are near enough to those of Earth to be empathised with but which contain sufficient indication of a Golden Age to be powerfully attractive. This theme is continued in the account of the despoliation of the Shire: when Bilbo/Odysseus returns, the landscape described by Tolkien must be rather like any rural but industrialising part of Europe or North America in the first third of the nineteenth century.

Science fiction is another convention which requires a rather specialised attention to settings and surroundings. In our context there are two main kinds: those which rely largely on a plethora of very advanced technology to sustain their interest and beyond that contrive to find themselves either in rather recognisable environments, or at most in a totally artificial city of some kind. Caves too seem to be important. The other kind, seemingly the more creative, is the invention of a new ecology which has enough in common with that of Earth to provide a link of credibility but is at the same time sufficiently different to allow startlingly new perspectives. Frank

Herbert's *Dune* novels are probably the outstanding examples; the planet on which they are set is a desert geoid and the key element in short supply is water: when a death occurs then reclaiming the organism's water is all-important, recalling the reclamation of phosphorus in Aldous Huxley's *Brave New World*. Beyond the displays of imaginative technology, though, science fiction rarely carries a different environmental message from other popular fiction.

To discuss place in 'serious' novels is not to know where to begin, for in so many it has an important role, in creating the right atmosphere for the interplay of characters or in providing symbols appropriate to the story. In any discussion of the novel in this context, Thomas Hardy is bound to be mentioned since he creates a whole landscape of his imagination which is nevertheless recognisable place by place, with the new names pushing us all just one step further from everyday reality so that Hardy can tell stories which otherwise might not be somehow allowable. The crucial test of this comes in *Tess of the d'Urbervilles* (1891), where Tess especially seems to be in the grip of forces which are so great that they must be tied to mythic orgins (The President of the Immortals, no less, puts in an appearance at the very end) and to the forces of the earth itself. When things go well for Tess, then the environment shines and is lush; when things are awry then we are back to the blasted heathlands and the stark eeriness of Salisbury Plain and Stonehenge. (This contrast is rather lost in the Roman Polanski movie since it was shot largely in Brittany and the granite houses are clearly wrong.) In fact, Hardy draws our attention in the same novel to a crucial change that was to affect the whole basis of rural life: the advent of fossil fuel in the shape of the threshing machine powered by a steam engine. Late in the story (Chapter 47, out of 59), the engineman:

> was in the agricultural world, but not of it. He served fire and smoke; these denizens of the fields served vegetable, weather, frost, and sun. . . . The long strap which ran from the driving-wheel of his engine to the red thresher under the rick was the sole tie-line between agriculture and him.

The engineman is, too, in the service of 'his Plutonic master', and it is during this threshing that Alec d'Urberville turns up again and as we all know ends up rather spoiling Mrs Brooks's morning routine. So the intrusion of the new economy is a turning-point in several ways.

The theme of industry as an intrusion into a rural landscape which can very likely set things off on a downward path comes as late as the 1920s in D. H. Lawrence, for whom we might have thought this was no surprise. But in *Women in Love* (1921), for example, colliery-owner Gerald Crich is seen very much as a power, just as is the coal (he has 'a great reserve of energy'), but in the end a destructive one: indeed self-destructive, we are invited to surmise. Being an 'arch-god of earth' did him no good once he got among real mountains. Birkin, who in the 1960s would have founded a self-sufficient commune in Oregon or Pembrokeshire, is closer to the earth, somehow, and thus is a survivor. In this respect, he is cognate with the famous gamekeeper in *Lady Chatterley's Lover* (1928), whose success at fostering young life in the pheasant-pens is translated to Constance's emotions, and it is in farming that Connie and Mellors will find their future, not among the colliers, 'a sad lot, a deadened lot of men: dead to their women, dead to life'. For a modern evocation of the power of death over an individual and the way in which it can, harnessed to the seasons' change, be eventually transformed ('There was winter. There would be spring'), Susan Hill's *In the Springtime of the Year* (1974) is riveting in its focus upon the details of human and natural life.

The novel is a natural vehicle for exploring an Utopia, and many science fiction books are of this kind. But the mainstream ideas of today's environmentalists (with special emphasis on population control) found a well-known protagonist in Aldous Huxley when in 1962 he published his idea of Heaven in *Island*. It has never become famous like his idea of Hell (*Brave New World*, 1932), perhaps because it all seems so reasonable (the epigraph quotes Aristotle on avoiding impossibilities in framing ideals) and even human death from cancer is somehow dignified by the presence of people for whom the materialistic strivings of the West have been renounced. Positivism gets a bad time, LSD gives a good time, and when in the end evil triumphs ('Progress, Values, Oil, True spirituality') we feel not that the whole island dream has been a nonsense, but really quite angry about this outcome. So we can at this level be swept up into an alternative world by a really creative writer.

The identification of humanity with place, however, is rarely as complete as it is in the major works of James Joyce (1882–1941), and in particular *Dubliners* (1914), *Ulysses* (1922) and *Finnegans Wake* (1939). Although Ireland features in all of them at a first level, they were all written in other places (Trieste, Zurich, Paris) and so we need not be surprised by any cosmopolitanism of approach that we may find, especially since Joyce was fluent in the main European languages. Even in the stories in *Dubliners*, which are written with a conventional use of language, the progression from youth to maturity, and the circularity of beginning and ending with a death as a prominent motif, there is a movement towards the unity of people and their world. At the end of *The Dead*, Gabriel Conroy's consciousness of the meaning of love and death is gathered up in 'snow falling faintly through the universe and faintly falling . . . upon all the living and the dead'.

In *Ulysses*, Joyce wanted to present a picture of Dublin, 'so complete that if the city one day suddenly disappeared from the earth it could be reconstructed out of my book'. So perhaps it could be, but the identifications are far wider, for the story of Leopold Bloom's wanderings on the one day in June 1904 connect outward to the rest of the known world just as did those of Odysseus in the Mediterranean. *Ulysses* can be seen as a multiterritorial pun and the spaces of the Odyssey can be mapped on to Joyce's Dublin. If we believe, as did Joyce, that it is possible to reconstruct the routes taken from Ithaca as far as Gibraltar, then they can be put also on to maps of Dublin as with a reverse pantograph. And if the Homeric epic is at heart derived from the Greek experience in colonising the western Mediterranean, then coming home is the key element in both, with time being used to extend space in *Ulysses*. Outwards again, the connection with myth is stronger than it first appears since many heroic figures like Apollo, Hercules, Odin and Finn MacCool are migratory, following the sun to its sinking which in midsummer in Ireland is in the northwest. Similarly, Molly Bloom's bed has travelled 'all the way from Gibraltar' and it is here that Bloom finally comes to rest ('He has travelled'), though for him the fixity of the bed and what it represents is scarcely so solid as that of Ulysses, whose furniture was built, we may remember, around an olive-tree. The identification of a person with the cosmos is made explicit in Molly's gynomorphism, in which the cardinal directions are assimilated to parts of her anatomy, and in which her sexual organs are west-facing and cognate with the word 'yes'. So in the final parts of the famous interior monologue which ends *Ulysses*, the homecoming and the myth and the poor imperfect humans all fuse into one affirmatory yes.

Finnegans Wake takes us a step or six further, with an even more experimental and apparently stretched form of language, in which the multiple pun, with no language excluded, makes it a kind of cosmic crossword puzzle. Here, Joyce seems to have compounded a total world, with many chthonic links: characters are often described in terms of what they eat, for example. Any critical study of *Finnegans Wake* is to some extent a gazetteer which must identify and signify the many place-names which are thick upon the ground of the text. The assimilation of the large cast of characters in *Finnegans Wake* with wider concerns comes at once in his use of Giambattista Vico's theory of history which is cyclic: societies are theocratic, aristocratic, democratic and then a *ricorso* brings back the gods. So all is linked to this cyclic nature of history, just as the river Liffey in various nomenclatural disguises is itself cyclic, being reborn from the clouds which come off the sea in which the river has in effect died. And the main female character, Anna Livia Plurabelle, is gynotropically assimilated to the river: she absorbs space as it were. So in *Finnegans Wake*, Joyce seems to have moved out from Ireland and become universal: the people somehow seem smaller than in *Ulysses* and the cosmic linkages vastly more important; if he had lived longer what might he have imagined next?

We could go on multiplying examples without perhaps doing more than showing that in the English language many writers of fiction have taken the non-human world very seriously. At the extreme of their imagination, however, they seem to have abandoned any duality of 'man' and 'environment', and we shall need to search for this theme in the greater intensity of expression which characterises poetry. Before that, though, we need to be reminded that writers in English and 'homed' in the British Isles have no monopoly on this kind of expression. Here also, some writers use place as a background, others as symbol, as an all-encompassing metaphor, and a few even try to provide a literary equivalent of the warnings inherent in some versions of the Gaia hypothesis. Such is the theme (not perhaps the right word for a multi-skeined narrative) of Gunter Grass's *The Rat* (1986).[58] This apocalyptic novel interweaves an acid rain version of Hansel and Gretel, a birthday party for an old Kashubian woman, the supplanting of humankind by rats after the nuclear holocaust and some bogus wall-paintings in the Mariankirche in Lübeck into a questioning of consumerism and the abrogation of control over our lives and minds that appears to be part of that particular Faustian bargain.[59] The central character, a female brown rat, tolls the bell on the penultimate page:

> I see multiplying nations. Free from humans at last, the earth provides space. Again, there will be plenty of fish in the sea. On the hills behind the city there will be dense forests. Birds will make use of the sky. New, undreamed-of animals will appear, among them mammalian bluebottles. But the old Danzig is falling apart. The richly ornamented façades are crumbling. Towers are falling asunder.

Grass will have to stand proxy for all the European writers who have been affected by the Green and Deep Ecology positions, just as one Japanese novel will have to serve to point up the sensitivity of this medium in that particular culture. Kawabata Yasunari's *Snow Country*[60] won the Nobel Prize for Literature in 1986 and is set in North-west Honshu, beyond the mountains from metropolitan Japan, where the snow lies deep and long, and where men with emotions frozen by culture and the city can learn that they and not their surroundings have to change if their humanity is to be realised. Shimamura, the protagonist, is in the train going to the mountain resort:

> In the depths of the mirror the evening landscape moved by, the mirror and the reflected figures like motion pictures superimposed one on the other. The figures and background were unrelated, and yet the figures, transparent and intangible, and the background, dim in the gathering darkness, melted together into a sort of symbolic world not of this world.

On another visit, Shimamura is in the resort in autumn, when the insects start to die:

> A bee . . . walked a little and collapsed. It was a quiet death that came with the change of seasons. . . . For such a tiny death, the empty eight-mat room seemed enormous.[61]

100

So here in Japan before the Pacific War, nature and humans never quite melt together in reality; afterwards in the city of Kamakura as portrayed in another famous novel of Kawabata, *The Sound of the Mountain* (1970), nature puts down tendrils from the hillside towards the house where most of the action takes place, but is normally kept well at bay: western dualism has come to stay even though General MacArthur seems to have gone home.

One trend that can often be observed in the novel as we move from the nineteenth century towards the present is the distillation of scene and occurrence. The sprawling and leisurely pace of Dickens or Scott, for instance, are gone and there is an intensity of writing, influenced perhaps by the added visual qualities we now expect when our imagination (used as it is to films, photography and TV) is appealed to. Intensity in words, however, is the realm of poetry: a tree born in a land without soil.[62]

Poetry

Formal definitions of this art focus on the technical questions of verse form and metre; a more functional view would perhaps concentrate rather on the way it echoes its Greek roots in *poiesis*, a making, a creativity, which Heidegger transforms into 'a bringing-forth'. In this way, we can see the poet's endeavours to use language creatively and effectively, and for its own sake. The cognitive world of the writer is not then like a conventional photograph but a fresh way of seeing the world: it lights it up clearly and forcefully since good poetry has a special intensity to it. Even my own limited knowledge of the literature presents enormous problems of selection. So a few authors will be presented in rather short order before a more extended treatment (still not very long) of T. S. Eliot (1888–1965), which parallels the singling-out of Monet and Joyce in the earlier descriptions of painting and prose.

One of the outstanding nature poets is John Clare (1793–1864), whose poetry takes in the everyday nature of the countryside of the English midlands. 'Takes in' is used because there is such an immersion in, and identification with, the plants and animals that we almost feel Clare to be a model for a drawing of The Green Man of folk legend. Because of madness, he probably never saw the economic changes which were coming on and which were to negate his view of nature and the seasons that 'Time loves them like a child and ever will.'

Despite some underlying tensions, the country was for him something of a paradise. Not so for George Crabbe (1754–1832) who lived in Aldeburgh on the coast of Suffolk. This region of the Sandlings was not very fertile and Crabbe wrote, much against the fashion of the times, of the toil needed to crop the area and also of the unromantic shorelines formed by long shingle beaches and mudflats. He lamented the state of the small farmer and the agricultural labourer and he also castigated large landowners who turned farmland over into park and 'improved' it in various ways:

> He filled the moat, he took the wall away,
> He thinned the park, and bade the view be gay.

His projection of the region into our consciousness today was secured by the use of part of 'The Borough' as the libretto for Benjamin Britten's opera *Peter Grimes* (1945); the orchestral interludes capture the atmosphere particularly well.

On the other hand, William Wordsworth (1770–1850) positively exuded a reverence for nature, to the point where in any anthology of landscape or nature poetry in English he would likely be given pride of place. As a Romantic, he found much of his material in the tension between the theme of nature and the primacy of the individual's imagination in the face of the dehumanising forces of the oncoming industrial revolution.[63] ('Little we see in nature that is ours.') His honouring of nature was such that the unsympathetic William Blake is said to have blamed some of Wordsworth's poems for a bowel complaint that nearly killed him. In essence, Wordsworth seems to have worried that imagination and nature might always be sundered and that only some apocalyptic change would restore a lost integrity where he could find 'Paradise, and groves Elysian, . . . a simple produce of the common day'.

Here we are not far from some of the positions of today's Deep Ecology movement (p. 135) and even the milder position of low-impact lifestyle communities. Indeed, Jonathan Bate talks of a 'green reading' of Wordsworth which places the poet's respect for the earth and scepticism about economic growth (both squarely in the Romantic tradition) into the midst of today's concerns, and allies Wordsworth's history of radicalism to today's Green politics.[64]

A sterner interpreter of the rural scene and its changes is the Welsh poet R. S. Thomas (1913–). His earlier collections are full of the harsh life of the Welsh hill-farmer, for whom this is a marginal way of life sustained on subsidies, threatened by afforestation, by tourism and by wholesale buy-outs of property by urban English. The new order is symbolised by 'Cyndyllan on a tractor':

> Gone the old look that yoked him to the soil;
> He's a new man now, part of the machine,
> His nerves of metal and his blood oil.[65]

This somehow completes the trend started by the engineman in *Tess*. The bleaker side of the modern holiday industry[66] is put forward in a most un-brochure like statement:

> We've nothing to offer you, no deserts
> Except the waste of thought
> Forming from mind erosion;

And it gets grimmer in later collections, through the loss of farms and the oncoming tide of spruce trees, and indeed religious doubts which must come very hard upon a country priest.

Such themes are intensified when urbanisation is considered. Philip Larkin, in 'Going, going'[67] laments the passing of fields and farms for:

> More houses, more parking allowed
> More caravan sites, more pay

to the end-point that:

> . . . greeds
> And garbage are too thick strewn
> To be swept up now

a pessimism which, although perhaps directed at the superficialities of environmental change, also hints at underlying western demands upon environmental resources.

The year 1922 saw the publication not only of James Joyce's *Ulysses* but of T. S. Eliot's *The Waste Land*. At the time both created a stir, though neither would do so now. Both are more mythical than our previous examples. *The Waste Land* is a long narrative by a *persona* who has something in common with Eliot, though it cannot be said to be directly autobiographical. The critic Grover Smith[68] says that spatial form is not a useful way to see the poem (although there are discrete scenes and pictures which are of that kind), and that it all resembles a dream or a symbolist landscape. So quasi-isolated vignettes of place and surroundings are important in *The Waste Land*: London recurs more than once. Any symbolism of fertility that the River Thames might offer is sullied by the contamination of the waters: the negative in the sentence is surely ironic:

> The river bears no empty bottles, sandwich papers,
> Silk handkerchiefs, cardboard boxes, cigarette ends
> Or other testimony of summer nights. (175–7)

> The river sweats
> Oil and tar (266–7)

London is only redeemed, it seems, by the 'Inexplicable splendour of Ionian white and gold' afforded by the church of St Magnus Martyr in Lower Thames Street.

Beyond this, and the grinding aridity of the desert in 'What the Thunder said' (331–58), there is the mythic affinity of the poem with the grand cycles of winter and spring and of death and rebirth, codified with the help of the Grail Legend and *The Golden Bough*. There is a reversal of some of the normal progressions: April here is the cruellest month, quite unlike the hope which Chaucer, for example, gives it. In this poem we have the sprouting of corpses. These myths point to an exploration of disintegration (critics seem to vary as to what is falling apart and how much), with the modern city very much the symbol of the process:

> Falling towers
> Jerusalem Athens Alexandria
> Vienna London
> Unreal   (373–6)

The end is only partially hopeful in its resolution: the myths of death and rebirth (including perhaps those of time in the cosmogonic sense) held less certainty than ever for 1920s people.

Eliot's masterwork of poetry is generally acknowledged to be the *Four Quartets* of 1936–43. They attract our attention immediately by their names, which are those of places which had some significance for Eliot and in the case of Little Gidding for others as well. There is some topography in the poem: London once again occurs, as in the tube and after an air raid in World War II, and there is a vivid snapshot of the road from Yeovil (Somerset) into East Coker:

> And the deep lane insists on the direction
> Into the village, in the electric heat
> Hypnotised. (*East Coker* I)[69]

However, there is more significantly a sense of the interruption of the orderly progress of time in *Little Gidding* where a season outside the normal succession occurs, when snow gives the appearance of the blossoms to come but 'Not in the scheme of generation'. Later, the works of mankind too are clearly transitory:

> Water and fire succeed
> The town, the pasture and the weed.

> Water and fire shall rot
> The marred foundations we forgot,
> Of sanctuary and choir. (*Little Gidding* I)

which seems to reverse the rather confident rhythms and orderly changes of the beginning of *East Coker*:

> . . . In succession
> Houses rise and fall, crumble, are extended,
> Are removed, destroyed, restored, or in their place
> Is an open field, or a factory, or a by-pass.

> . . . there is a time for building
> And a time for living and for generation
> And a time for wind to shake the loosened pane

Thus when Traversi[70] thinks that the links which unite man and nature have fallen apart, and this causes ruin, we can see the regression from one state to another and less desirable one. The resolution for Eliot is only to be found in the stillness of his religious faith.

The work of T. S. Eliot is often claimed by US critics as American, and his work can serve as a bridge to the writing of US nationals. In our framework, there are many with claims to be considered but let us select just three of particular interest: Walt Whitman (1819–1892), Robinson Jeffers (1887–1962) and Gary Snyder (1930–). Whitman is apt to come over to us now as one of the larger-than-life types in the Hemingway and Miller mould and there is no doubt of the heroic stance of much of his verse. There is, though, a sensitivity (in e.g. 'When lilacs last in the dooryard bloom'd' and 'Out of the cradle endlessly rocking') which they would have found somewhat foreign. Whitman's 'Song of Myself' is one of his great incantations in which he identifies with everything in a non-dualistic way:

> For every atom belonging to me as good belongs to you

and later:

> I think I could turn and live with animals, they are so placid and self-contained, . . .
> They bring me tokens of myself . . .

which is a development of:

> I believe a leaf of grass is no less than the journeywork of
> the stars,

and ending:

> I bequeath myself to the dirt to grow from the grass I love,
> If you want me again look for me under your boot soles.

So a closer identification with fellow humans and their activities as well as with nature can hardly be imagined: indeed in his refusal to differentiate between different forms of human activity, Whitman is perhaps open to charges of being too welcoming to events like 'the quadroon girl is sold at the auction stand'.

Robinson Jeffers went further: his general thrust was towards the position that humans were second-class compared with nature: he labelled his outlook inhumanism. The raw material of his poetry is the elemental emotional state of man and his encounters with the landscapes of the West of the USA: he was through and through a Californian. He had a gift for a telling phrase or comparison which linked humanity and nature: 'humanity is the atom to be split' is one of the most famous; there is also the likening of human progress to that of the spawning salmon: 'To find its appointed high place and perish', and the comparison of Judas:

> . . . you enter
> his kingdom with him, as the hawk's lice with the
> hawk
> Climb the blue towers of the sky under the down of
> the feathers.

His urging to love all of nature, 'Not man apart from that', led to the title of the Sierra Club newsletter. His idea of Inhumanism is primarily poetical rather than logical: it emphasises a shift of emphasis from man to not-man but at the same time suggests this detachment has human value; the whole package offers no difficulties to modern Californians after their contact with Zen, nor to those whose ethical systems reach out to a view of nature which possesses intrinsic rather than instrumental value: he praises the weasel, the hawk and the heron because:

> These live their felt natures: they know their norm
> And live it to the brim; they understand life.

In some ways, therefore, there is a mysticism like Whitman's which calls for identification with all men and women regardless of society's evaluation of them and with nature. We can recognise the resurgence of this type of view (or at least of part of it) in some of the environmentalist philosophies discussed in Chapter 5. One critic[71] suggests that Jeffers' Inhumanism leads to a passive withdrawal from the world of human history: if so, then Jeffers is for present readers an interesting signpost but on the road to nowhere.

The influence of non-duality of humans and nature via the thought or practices of Zen Buddhism in the West of North America is best illustrated in poetry by the work of Gary Snyder. His main work dates from 1973 onwards, as a reaction to the instrumentalism of the UN 1972 Stockholm Conference on the human environment. It is a deeply felt, openly informative body of work founded on a non-duality of humanity and nature. It aims at a kind of wisdom which is policy and prescription as well as description and prediction. The unity of all is founded on homopiety (a conviviality among humans) and geopiety (the connaturality of humans with the whole of nature) and the whole is labelled 'ecopiety'.[72] It is strongly influenced by Zen Buddhism and refuses to equate 'to be' with 'to have' and celebrates all things 'wild, sacred and good':

> hummingbird
>
> > intelligences
> > directing destructing instructing; us all
> > as through music:
> > songs filling the sky.
>
> Air, fire, water and
>
> > Earth is our dancing place now.

Whether in fact this represents another lament for a lost Golden Age or whether it is the viable seed of a future vision of something entirely different, has yet to be seen.

The landscape in this poetry is not simply there: it has a symbolic function that has been there since early days. Cities have represented the ideal commonwealth ever since the Pilgrim Fathers; islands are places of greenness and dancing; mountains represent a challenge as well as the potential of wealth.[73] The environment then is full of meaning in this poetry and moreover it is what Auden called a *paysage moralisé*; among other things it urges us to rebuild the city rather than dream of islands: be cornucopian rather than catastrophist.

Outside the West, Asia is a deep well of poetry: the work of the Chinese became the best known abroad in the first instance, followed by that of Japan and the Indian subcontinent. Even more than that of the European languages, it loses its flavour in translation: the quality of lapidary terseness in particular comes across badly in tongues like English given to the insertion of 'the', 'an' and 'hers' at frequent intervals. The oriental poetry now most familiar to us is perhaps the short Japanese verse known as the *haiku* (a three-line form of five, seven and five syllables) in which the Buddhist concepts of non-duality have often been expressed, especially by classical masters such as Bashō (1644–94), Buson (1716–83) and Issa (1763–1827). The *haiku* should contain an allusion to the season, and preferably an emotional ejaculation, and so the blending of human life and nature is foreshadowed in the form itself. The *haiku* has proved itself adaptable to modern writers but the collections such as Bashō's prose account[74] of his *Narrow Road to a Far Province* (undertaken in 1689) which is studded with *haiku* and other verses give the flavour best, as in:

> A mound of summer grass:
> Are warriors' heroic deeds
> Only dreams that pass?

and:

> Loath to let spring go
> Birds cry and even fishes'
> Eyes are wet with tears.

which is not far from:

> In the fire, a telegraph pole
> At the heart of the fire
> A telegraph pole like a stamen
> Like a candle,
> Blazing up, like a molten
> Red stamen.

This is part of a poem[75] by Hara Tamiki (1905–51) who had first-hand experience of Hiroshima in 1945. There is much subtlety to Japanese poetry: even the human emotions can be unstated except in the alliteration of a melancholy letter like *k*; Bashō again:

Kare eda ni	On a bare branch
Karasu no tomarikeri	A rook roosts:
Aki no kure	Autumn dusk.

We could, perhaps ought, to go on: the pulses of nature and their connections to mankind appear time and again in most poetry. But in a short space, let us be convinced that these themes are likely to be there and to be found and appreciated (enjoyed, even) and do not need a Ph.D. in Literature: adapting Husserl, we might say 'to the texts themselves!'.

Journalism

By definition more ephemeral, nevertheless this form of writing may have considerable effect, since it may be precisely the type of information which decision-makers actually read. We often see a book review which says something like, 'all those concerned with making decisions about the environment should read this book', and sigh, knowing that they never will. But something shorter and more popular might just catch their attention, as well as those millions whose reading material does not extend beyond the tabloid press.

There are, of course, magazines devoted to almost every aspect of environmental matters, from the whole planet down to the *Snail-watchers' Gazette*. These are mostly one type in the sense that they do not question the dominant paradigm of the industrial society, unless they are from the 'alternative' culture. There were a great number of these latter in the 1960s and 1970s but relatively few have survived. *The Ecologist* is always alternative in its views but is relatively expensive and serious, without the element of extravagance in ideas and layout that the 'drop-out' culture produced; the same is true of *Resurgence*. Popular science nearly always includes environmental material but is confined to the scientific aspects, often in a rather artificial way; however, *New Scientist* is a compulsory read for anybody with such interests. Other magazines tend to latch on to the current worry (carbon dioxide, sea-level and the ozone layer at the time of writing), and then drop them assuming that they have thereby been solved.

The same is roughly true of the serious newspapers: they will take a current topic and give it more than two columns (especially on Sundays) in the hope that they will thereby influence the gate-keepers and decision-makers and so enhance their reputation as fearless prosecutors of the public interest. They may even run a special supplement for a while. Interestingly, it looks as if the actions taken by western governments in the late 1980s to reduce the production of CFCs came about as the result of scientific advice and the pressure from scientific meetings rather than after very much attention in the press, where holes in the ozone layer were more difficult to dramatise than e.g. the controversy over clubbing baby seals in Canada.

This brings us to the popular press. For them, only major events of direct concern are worth anything other than a brief mention. Animals are usually the focus; abstract matters like the sea-level or the ozone layer are difficult to translate into the terms which the editors judge their readers to be capable of comprehending. And there is never any questioning of the basic world-view: that too is deemed to be beyond their readers.

THE OTHER ARTS

This section is necessarily a catch-all piece since it deals with some of the other manifestations which, while not insignificant, are nevertheless not major contributors to environmental constructions in the manner of the visual and literary arts. It seems scarcely possible to give a good account of the most important of them in popular terms, television, because the global offering is so diverse, but some extended mention must be made. Cookery, too, can be elevated to the status of an art for those lucky enough to have that sort of choice and it then makes an implicit environmental statement, though it is difficult to generalise on a world basis. So we may well have to venture into some uncharted waters in this section, in the company of a distinctly leaky boat.

Music

We can possibly think here of explicit and implicit music as far as nature and landscape are concerned. At one extreme, actual sounds may be incorporated or imitated as by Beethoven in his Pastoral Symphony's rendering of bird-calls, an idea much used by Messaien in this century. The title of a piece may be used to evoke place even though thereafter the music flows entirely from the composer's imagination. More common, though, is the use of regional idioms to evoke the scene: folk-music is particularly popular. Think, for example, of the numerous nineteenth and twentieth century pieces with the word *español* in the title and using one of the characteristic rhythms of Iberian folk-music. Modern composers trying to escape these conventions often face problems in finding the right response in their audience and have often to resort to setting words which are evocative (like regionally based poems) or an explicatory programme note.

The depiction of landscape is perhaps most advanced in the work of the US composer and insurance executive Charles Ives (1874–1954). Some of his work is overt in title and references: *Three Places in New England*

(1908–14) uses American hymn and marching tunes, for example. But Ives tried to get beyond that kind of structure to a dialectic in which the mind of the listener has to grapple with more than one idea at a time. Ives likens this to looking at a landscape:

> trying out a parallel way of listening to music, suggested by looking at a view (1) with the eyes towards the sky or the tops of the trees, taking in the earth or foreground subjectively – that is, not focusing the eye on it – (2) then looking at the earth and land, and seeing the sky and the top of the foreground subjectively. In other words, giving a musical piece in two parts, but played at the same time.[76]

The result is Ives's unfinished *Universe* symphony (1911–28), which needs to be heard[77] to judge whether this translation of a basically visual way of environmental cognition to a sound-sequence can be successfully made. If it can, then composers can try not merely to depict the basically static elements of a scene but to think about the processes as well, even without using words.[78]

At the more implicit level there are conventions, too. The use of the French horn (better, several of them) in the incidental music for Westerns is a commonplace, as is the type of surging on strings and brass if the sea is involved; and quiet rippling on woodwind over thrumming strings is a must for pastoral countryside. If a recording of the *samisen* is not available, then queer chords on plucked strings will tell us that we are east of Eden. Sacred music will often carry its message out of the spaces for which it was composed: the organ music of J. S. Bach, for example, must carry for most listeners a sense of the numinous even when played in the starkest modern concert-hall and thus change the lifeworld of the individual hearers if only temporarily.

Historically, R. Murray Schaefer[79] tells us that when the natural sound-scape was overrun, then a whole load of sensitive reactions in music was created, reasserting the natural or the delicate. He also quotes Ives's *Universe* symphony, but along with the work of Olivier Messaien and Claude Debussy as examples of a tender reaction to industrialisation. We might compare Debussy with Monet, perhaps. The influence of a changed world upon musical forms can also be seen, argues Schaefer, in the way the sounds of the railroad (such as wheels over track ends) influenced jazz. He also suggests that the advent of recording was relevant in the development of the short athematic pieces we associate first with the names of Schoenberg and Webern.

There does not seem, though, to be much 'Green' music. One example is the *Missa Gaia* by the US composer and saxophone player Paul Winter. This uses an orchestra, cathedral choir, solo soprano saxophone, and recorded sounds of whale and wolf to intersperse some of the sections of the Latin Mass with hymns of a modern type and Franciscan leanings. Curiously, it lacks the Credo or even any equivalent ('I believe in one ecosystem . . .'), and it does not seem to have spawned a great number of other such compositions: the true vehicle of Green music is still the informal folk-song, though some pop groups have ridden this bandwagon without seeing any irony in the amplification and lightshow equipment. In the field of 'classical' music, then, we must perhaps look for implicit relations with nature rather than any overt statements: the symphonies and tone-poems of Jan Sibelius (1865–1957) might be a good place to start.

We might suspect that all this discussion gets us virtually nowhere and that the real place of music in environmental constructions is truly pheno-menological in terms of the meaning which a tune or a whole piece has for one person in terms of one time. In rather trite words, 'they're playing our tune' sums it up. More general statements in our present field of interest (though not in others: think of the uses of military and patriotic music) are probably very uncertain in their validity.

Cinema

During the lifetime of the talking motion picture, its social appeal has changed more than once. Until the advent of universal television in the 1950s, it was a popular medium, in tabloid format as it were. After the success of TV it became more of an art form in the West, though retaining its popularity in nations with restricted access to television and in France. There is currently something of a revival in cinema-going in the younger age-group, and the making of full-length films has found a new outlet in video. So while we are not dealing with a mass medium any more, there is no doubt that any attitudes to environment struck in a film will reach quite a large audience. Should we discuss the 'typical' or the famous? At the risk of being over-selective, let us try for both. In the West, for example, the films of Akiro Kurosawa are 'art' movies, in Japan they are much more popular, for obvious reasons. If we take an example like *The Seven Samurai* which is probably his best-known work, then we can read into it a very distinctly Japanese attitude to the natural world, as well as the social order. The non-duality of humanity and nature which is central to traditional Japanese thought in this field is evident in, for example, the way all the characters in the final battle (which is everybody who matters) are merged into one mass of earth and water by the incessant heavy rain. All this is emphasised by the making of the film in black and white. Then again, a major character meets his death in the river, making another identification with the forces of nature. It is also possible to read this death as a harbinger of all kinds of change since it comes from a gun, not from sword, spear or arrow. But if it really adumbrates ecological change then it is still 200 years to go to the industrial revolution in Japan: a social levelling seems more likely.

111

Probably the most 'environmental' Japanese movie of recent times is Kurosawa's *Dreams* of 1990. This has eight separate episodes, through which themes of nature, humanity and destruction are dominant. A boy follows a girl to an orchard in which eventually all the trees are cut down except one. Do we have to wait, Kurosawa seems to be asking, until there is only one peach tree left, before we cry? More overtly is the episode 'Mt Fuji in Red' which depicts the melt-down of Mt Fuji following some kind of nuclear accident. The film ends with an idyll in a water-mill powered village in which the death of the elderly is celebrated with dancing as an organic event. It is ethereal and beautiful but part of a Golden Age view of the world which is unrealistic, even seen in the context of Japan's love-hate relationship with western culture.

Conflict in a non-urban, non-industrial setting is at the heart of the batch of movies made about the involvement of the USA in Vietnam, Cambodia (now Kampuchea) and Laos during the Johnson and Nixon presidencies. The films were mostly made in the 1980s (and usually in the Phillippines) and clearly form part of a national effort to come to terms with these years: there are analogous films in West Germany dealing with the *Hitlerzeit*, especially those of Rainer Fassbinder. The Vietnam films hark back to a virtually medieval European view of the forest ('jungle': always a resonant word) as a hostile wilderness containing the agents of evil and death; which latter it did for the G.I.s. There is something of *Beowulf* in many of these films and it is not surprising that one weapon deployed was that of defoliation, as well as practically any other method available, to remove trees. There is a long tradition of using fire in forest-located warfare and here, of course, napalm is added to the armoury, though not for the first time.[80] The aversion to the natural world, including the cultivated land of the peasants, is emphasised by the dominance of the helicopter as a means of transport. It is difficult to imagine a device for more rapid separation from the land, and which in this context symbolises all kinds of connections: with the comforts of base and of Saigon for the living, and burial back home for the dead. The most thoughtful of these movies is Coppola's *Apocalypse Now*, which is based more or less on Conrad's 1902 novel, *Heart of Darkness*. The successful penetration to the kingdom of the renegade soldier is made in this case by boat and perhaps this threading of a maze by using a feature of the natural environment like a river is a clue to the 'success' of the mission.

The ultimate 'place' movie so far is probably the 14 hours of Edgar Reisz's *Heimat* (1984). The word is full of overtones: first of all it is redolent of the regionalised setting to which the characters all belong and whose dialect they speak, in this case the Hunsrück. But it resonates with the Nazi ideology of *lebensraum* and also of the plight of the dispossessed after 1945. The role of place is crucial: getting away from the land is the mark of success: whether to the USA or to selling clocks in the town or even putting up a mock-mansion in the village. To stay rooted in the village, as does Maria, is to court all kinds of failure and loss. There is not much romance about the Hunsrück countryside: to that extent the film is a faithful reflection of what most farming communities in Europe feel about their immediate environment: can it add to the bank balance?

There are also explicit nature movies. Their origin is probably in Disney animations but they use the greater reality of location film either to tell a story of nature ('George the Grizzly') or a human story set in the wild ('The Malton Kids Go Backpacking'). These are often shown on TV at holiday times when programming is difficult to fill up and are somewhat predictable: in attitudes they reflect most of all the 'shallow ecology' school of North American environmentalism: the wild must at all costs be preserved from development but without giving up any of the modern conveniences like rescue helicopters or the basic matrix of an industrial lifestyle. Because there is a happy ending and since the wild is shown to be basically beneficient to mankind, then space is substituted for time in the reformulation of a myth of a Golden Age.

Television

The mix of programmes received in any one place is such that comment other than from personal experience is very difficult. The very diversity of offerings, getting more so as cable and satellite are more used, defies simple analysis.

The most obvious programmes are those documentaries devoted to almost every aspect of nature, conservation and (usually) degradational environmental change. These cover virtually any topic from yet another coral reef to alleged leukaemia clusters near atomic energy centres, and are generally of the format with a narrator (in voice-over) and interviews with various people. In most instances, nature and the environment are not strongly linked with a 'personality' in the way many other TV shows are, though the UK examples of David Bellamy and David Attenborough are highly not-able exceptions. Generally speaking, these programmes are visual journalism in the sense that they focus on today's worries as perceived by their producers (actually, next week's worries, given the lead times for production) and there seems to be little Green or Deep Ecology material: ameliorating the effects of the bad guys seems to be the underlying theme. Perhaps the environment is today's Western.

What about the ads? It is commonplace to hear it said that they are much better quality than the programmes and no doubt more dollars per minute are spent on them. What messages about the planet itself are sent out along with the insistence that we buy this product? Nature, it seems, is generally a benign place. It rarely threatens and more often is an appropriate setting in which values are conferred: an exotic setting presumably says that you too will be exotic if you use our shampoo. Occasionally less than desirable places (usually very dry or very hot and sticky) are presented and we are invited to assume that we will smell like them unless we use a particular deodorant. Now and again, nature presents a challenge like fire or rugged terrain and this has to be overcome with perhaps a new model of car, or cans of specially strong beer.

Behind all this is the constant background of depiction which may or may not be noticed by viewers. Just as there are associations which count every swear-word uttered and every incident of violence, there perhaps ought to be somebody counting all the times that nature is destroyed, or an environmentally degradational product is used. For the importance of TV cannot be gainsaid. The values of the West certainly and of many other cultures probably are now validated by this medium, just as they were formerly by print and before that by the Church. Activists ought to be keeping an eye on it.

Gardens

In pursuit of their two-fold purpose of beauty and productivity, gardeners effect a very intense transformation of their surroundings. The input of materials is very high: energy, fertilisers, biocides, machinery, domesticated plants are all used. This is paralleled by a very high cultural input in which the piece of ground is made over in the image which the gardener has acquired. Seen in this light, there seem to be two main kinds of garden, the first of which is the creation of some nature in what would otherwise be a totally built setting: no matter that the nature is as artificial as the rest of the place. Hence the Japanese garden no larger than a paving stone, the rooftop gardens and the window box. The second type is the manipulation of a less intensively used environment in order to demonstrate the appro-priate values for the users' pleasure-times. The formal garden of the great houses of Europe (and its watered-down versions still present in suburbia and beyond) demonstrates cultural *mores* rather well in the eyes of many commentators. The formal lines of the Italian gardens which come to their apogee in André le Nôtre's great park at Versailles can be read as interpret-ing the rigidities of a Copernican world in which the very stars knew their places, let alone the people; the Classicism of Haydn and Mozart says the same thing. The reaction of Rousseau to this saw, also at Versailles, the rather silly *hameau* built for Marie Antoinette where she could play at peasants and dairymaids; perhaps she learned how to cook *brioche* there. Romantic and heroic ideas produced, for example, the English landscape garden, with its informal clumps and winding lines, none of which was a whit more natural than its predecessors but satisfied the whims of those who could afford it. Today's gardens are more difficult to interpret but the substitution of the patio for the parterre, and the lawn plus rose bushes for the rest of the park seems feasible.

In the westernised cities of the world the public park has an important role: the lungs of the city, they are often called, without a very strict interpretation of the analogy. More likely, they are something to remind us of the wild that has now vanished but which is essentially well under our control: rather like seeing a wolf at the zoo.

Architecture

At first sight, architecture is the very antithesis of all that we have meant by environment since it is an attempt to control our surroundings totally and hence to replace what nature or the previous generation of humans provided: 'the built environment' is a phrase often used. It has not always been the case that structures ignored their surroundings: most low-technology building has to conform and adapt in terms of materials and orientation, for example, as well as to cultural *mores* and it has often been very successful in adapting to flood regimes or intense solar radiation or the need for nomadism, for example.

Freedom from the limitations of labour supply has been important to human societies from early times, however. We see it first in ceremonial buildings, which are so much bigger than all others, in Egypt, Mesopotamia, Greece and China, for instance. They are not always totally divorced from environmental considerations since their astronomical alignments might be very important; they might themselves be microcosms and therefore have to reflect the lines of the heavens, just as Christian churches have a liturgical East even if they are aligned some other way according to the compass.

Architecture after the industrial revolution, however, positively celebrates its release from the forces of nature. The skyscraper building is the obvious example of this: it grows as if gravity itself were no longer important, and its rectangular lines are those of the draughtsman's instruments rather than those of the natural world. Naturally there have been reactions to this: houses which grow over streams or incorporate trees, like some of Frank Lloyd Wright's buildings, are obvious samples, though they are not very numerous. In most large cities, the weather and the stars are now seen as virtually irrelevant except as they might pose heating and lighting problems; any desire for greenery can be satisfied by easy-upkeep plastic palms. A few mystics and visionaries have tried to devise architectural systems which relink us with the cosmos, like Paolo Soleri's arcology,[81] but a look out of the window of your plane as it descends over the suburbs will disclose a rather different world-view entirely. In other words, the construction made by architecture, both physical and symbolic, is that of the high-energy industrial world, which is the outcome of centuries of adherence to the idea that nature had to be overcome.

ACROSS BOUNDARIES

This examination of the treatment by various creative artists of the theme of humanity in nature has two obvious defects. The one is the relationship of these arts one with another, and the other is their relationship to changes in the overall relationship of culture with nature. There has been discussion, for example, that while some visual artists imply that life is comprehensible and controllable, others intimate a desire for change. A strong case can be made, and often is, that the 'modern' movement in the arts mirrors the disintegration of a viable relationship between humanity and nature.[82] *The Waste Land* is an epitome of this but many other examples of fragmentation can be seen in the written and visual arts, especially perhaps in the abstract movements, and in some of the more disjointed music of the post-Western period. Equally, the slab-sided high-rise building expresses a more or less total indifference to the world outside it. With postmodernism, on this reading, then the arts become further segmented into instantly saleable pieces, each famous for 15 minutes, to paraphrase Andy Warhol. Icons of the postmodern like the movies *Blade Runner* (1982) and *Wings of Desire* (1987) may possibly be designed to show us that only within ourselves can we generate a different, warmer, world-view.[83] No outside agency is, in the end, going to protect us from ourselves. In this view, the arts may be echoing what Wilfred Owen said of the poets' role in war, which was to warn.

THOUGHTS INTO ACTION

This chapter begs us to consider Nietszche's uncomfortable apophthegm that 'Art is more powerful than knowledge, because art is life while knowledge invariably ends in destruction'. Given as he was to drama, Friedrich Nietszche (1844–1900) invites us, in our present context, to consider whether the ways of natural science are inevitably degrading while those of art (including, we might hope, the art of living) are always constructive and evolving. This is clearly nonsense, even if heroic nonsense. Science can lead to the type of knowledge which enhances the world, just as the arts can be part of a downgraded style of life. Judging between what is creative in the longer run and what is destructive is, however, very difficult. Nevertheless it is a task which students of ethics have always set themselves and which lawyers have never shrunk from enshrining in black letters. We must now see what they make of our interaction with the non-human parts of the planet.

5

NORMATIVE BEHAVIOUR

Normative behaviour is concerned with rules, recommendations and proposals, and thus contrasts with mere descriptions and statements of an
external world. The use of the term generally implies that standards or values
are involved. By values we mean the moral principles and beliefs or accepted
standards of a person or social group, so that they are likely to be culturally
relative. But by moral we mean the assessment of human conduct in a given
situation: is it right or wrong and can it in fact be judged by some absolute
standard rather than be culturally relative?

A very little thought reminds us of the existence of the idea that we ought
to behave towards the environment in some particular way, a notion that is
usually extended towards our use of resources as well as the various other
kinds of environmental impact. Since different cultures impose different
outlooks, it will be necessary to see if there are any human universals which
can be applied here. That is to say, genuine human traits resulting from the
commonality of being human, rather than, for example, the *de facto* dominance
of the western world-view over so much of the inhabited part of the planet.

The abstract and philosophical side of this discussion centres around the
study of environmental ethics. Ethics is that branch of philosophy which
investigates morality: the varieties of thinking by which human conduct is
guided and may be appraised. It looks at the meaning, therefore, of
statements about the rightness or wrongness of actions; at motives; at blame;
and fundamentally at the notion of good and bad. There is often, too, a
question about the rightness and primacy of ends and means: if a region
seems to be over-populated in relation to available resources, is it right to
use any means to reduce the birth rate or must the means themselves be
evaluated for their morality as well? Ethics do not exist in a vacuum: some
ideas seem to grow out of the findings of other types of knowledge, as we
have seen with the ideas of ecological science.

The practical and institutional side of the material deals largely with
environmental law, plus some discussion of policies which flow from
legislation. Most nations (and often their regional and local governments)
now have bodies of law relating to environmental matters such as the

117

assessment of impact or the control of contamination, and there is a well-established field of academic study. Policy is not always easily disentangled from law, but is in general concerned more with the day-to-day implementation of whatever actions result from or respond to the enactment of legal measures.

Naturally there are unsolved problems of both ethics and law. One in particular is that of scale. Since so many ethical precepts have been developed in particular cultures, are they now outdated in a world which in many ways is drawing so much closer together? This is especially so in an environmental context, where many human actions can now have results at a global scale. National attitudes, let alone those at a smaller scale, may lag behind the development of an appropriate ethical awareness. This is magnified when the formulation of international law, to deal with trans-boundary problems, becomes important. The working out of enforceable legal measures to deal with e.g. the resources of the oceans, the pollution of the Mediterranean or the problems of acid precipitation, have all occupied years of legal and political time recently, with the basic difficulty of fitting the cognitions of the nation-states to those of the problems. Even within a relatively compact unit like the EC, differing views on right and wrong prevail. For a simple example, there is the Italian attitude towards the killing of small birds during their migration (Figure 5.1) which contrasts with that of northern Europe; more complex has been the British attitude to water-borne contaminants which insists that they are all taken away and diluted by the seas surrounding the British Isles and so any discharge limits should be lower in the UK than on the mainland. When it comes to the involvement of the United Nations or any other world organisation tackling problems of global importance then the complexity of the formulation of law and its translation into practice can easily be imagined.

This chapter then covers a wide spread from the restricted world and occasionally dense language of the academic philosopher to the highly public and usually dense language of the lawmaker and civil servant. But even the latter must inevitably start from the basic tendency of humans to distinguish right and wrong in most situations, to which the environment is no exception.

ENVIRONMENTAL ETHICS

While nobody with any feeling for intellectual matters would want to dismiss the work of writers in the field of ethics, it is not an easy topic for the non-specialist to penetrate.[1] Two considerations strike the novice immediately: first, that the language is by nature abstract, with words such as value, duty, rights, freedom, responsibility being at the core of the vocabulary. No wonder that books and courses on the environment are heavier on ecology than ethics: any difficulties over the definition of trophic

Figure 5.1 In Italy in the 1980s, public meetings were necessary still to gain support for policies that outlawed the mass killing of small birds. A poster photographed by the author in Padova in 1987

levels are easier for non-specialists than the case for the legal rights of the Spitsbergen Palm.

Concerns and principles

The first questions which ethicists and philosophers find it necessary to tackle seem to be (a) can we talk of environmental ethics at all? and (b) is it possible to talk in the aggregate or must there be a break-down into subsets of concern?[2] Some begin with an **ontological** argument[3] which takes the form of asserting that it is the duty of humans to promote or preserve the existence of good. The environment, whether as beauty or resources, is part of that good and its existence is physically contingent upon the continued existence of its components and its history, neither of which humans ought to disrupt.[4]

Further consideration reveals that there are (at least) two possible meanings of 'environmental ethics' to be discussed. They are:

1 The idea of an ethic for the **use of the environment**, i.e. a position which starts empirically from where we are, accepting the dominant world-view that the Earth is a set of resources which humanity is free to employ, even if some of them are employed in their entirety as aesthetic and recreational resources rather than simply as materials. The words 'utilitarian' and 'instrumental' are often used of such an attitude.
2 The idea of an ethic **of the environment** in which the moral standing of the non-human entities of the cosmos are given equal value with the human species. There is a 'weak' version in which at the very least this standing must be extended to all conscious beings and some non-conscious entities as well.

The first of these is well established and can be encapsulated by the term 'wise use'; the science of ecology has been harnessed since the 1960s as a hitching-rail for a management ethic for the human use of the Earth.[5] But another abstract element in the area management ethics must be our duty to future generations of humans. As yet unborn, they have no voice in our current preoccupations.[6] Normative behaviour, then, addresses itself to how much we should worry about the welfare of those to come: should we refrain from using non-renewable resources (like fossil fuels) now so that this patrimony is not denied to our descendants? Or would we benefit later generations most by turning all these resources into knowledge of how to do without them?

The second viewpoint is the more difficult in both abstract and practical terms. The idea of intrinsic or inherent goodness (and hence of moral equality with humans) has rested primarily upon the presence of value independent of the presence of any conscious being: the value resides in the object itself and is not conferred upon it from 'outside', rather in the manner of an Honorary Degree.[7] For humans, then, the fitting attitude is one of admiring respect coupled with the realisation that the environment is not merely a means to human ends. The espousal of such an attitude would not have seemed strange in the Middle Ages, but has been largely submerged or dissolved since the Renaissance and the Enlightenment by the narrow focus of humans upon humans.

The current notion of inherent value assumes, however, the kind of distinction between subject and object that we associate with René Descartes. But one of the major consequences of the findings of quantum mechanics during the twentieth century has been that such a differentiation cannot be made. At the fundamentals of matter, what can be said about a particle in terms of its velocity and location are to some degree chosen for it by the observer: she or he may choose to know the particle's position definitely *or* its velocity definitely or both approximately. Location and velocity are, as Callicott puts it,[8] potential properties of an electron variously actualised in different experiments. Any attribution of value, therefore, has to be focused on neither the subjective nor the objective: if categories are needed, they must transcend the old dichotomies. Further, it can be argued that the universe consists of just one substance – spacetime – which is 'self-realising', and which must therefore be an ultimate source of value.[9] Above the underground rings of the particulate world, the extension of such ideas means perhaps that the essential unit of the world is the identification between self and world; the human self is a temporary knot in a web of life and non-life, rather as a particle seems to be a temporary manifestation of energy. So nature is intrinsically valuable to the same extent that the self is valuable.

None of these sets of ideas is without its critics. At one level, it can be argued that the aesthetics which, for example, motivate much environmental concern are not as fundamentally human as eating and drinking, and that identification with nature is simply a metaphor. Further, environmentalism can be seen as an ideological descendant of the Romanticism of the early nineteenth century and so is likely to be identified with reactionary politics.[10] This concern with political interpretations can be carried deeper into the structure of the language we use. For example, it may be that landscapes and species to which we attach value are expressions of cultural values: in North America, 'the wilderness' is said to be a repository of male and nationalistic traits. Even further, it is said that the current arguments about environmental ethics are incoherent because they use terms that only make sense in a system which has an agreed concept of human purpose and direction, a *telos*.[11] At present, such terms as rights, interests, utility and duty are all disguises for a determination to hold on to power. So the concept of rights (if it is to flow from a determination of intrinsic worth, for instance) is merely a fiction hovering above reality. It may, of course, be a useful fiction for promoting change in human behaviour but it carries some other possibilities for abandoning the debate over environmental ethics since environmental 'problems' can be seen as social problems, to be solved by social action, with appropriate contributions from existing social and political philosophies.

In initial summary, therefore, the main foci of discussion in environmental philosophy and ethics at present seem to be:

- must an environmental ethic be based on human values, interests and goods or the corresponding features of the non-human world?
- does non-human nature have value in itself (i.e. intrinsic value) or only as a source of satisfaction of human wants (i.e. instrumental value)?
- can moral concern be directed only towards individuals or can it be directed towards groups or categories such as ecological communities and ecosystems?

The attempt to develop a different relationship with the non-human world, on paper and in practice, is gathering pace rather than abating, so we shall have to see in a little more detail some of the ways in which it is developing.

Pragmatics

To illustrate one practical outgrowth of ethical thinking about the environment, consider the 'lifeboat ethics' associated with the North American biologist Garrett Hardin.[12] Looking at resource availability in the future and at population growth rates, Hardin likens the situation to a series of lifeboats. The rich countries are like boats with a moderate number of passengers on board, the poor countries are like overcrowded vessels. The poor continuously fall out of their boat and hope to be admitted to one of the less crowded boats. According to classical Christian or Marxist ethics, says Hardin, everybody should be allowed aboard. This would lead to complete justice and equally complete catastrophe. Hardin argues that to help the poor at all (via technology transfers or food aid programmes, for example) is to diminish the safety margin for the wealthy and to reduce the choices for future generations. The stark impact of this outlook is somewhat modified by Ehrlich's 'triage' proposals, in which some selected individuals would be helped, following the practice of battlefield military medicine.[13] In this, casualties are divided into three categories: those who will die no matter what is done for them; those who will live even if treatment is delayed; and those for whom treatment makes the difference between life and death. These latter might be admitted to the lifeboats. Both these proposals attracted the realistic and the hard-headed among international development and financial agencies, just as they have evoked opprobrium from those who see the ideas as 'anti-people', from those who argue that justice ought to be maximised before general well-being, that our duties to the present generation outweigh those to future generations and that democratic decision-making would produce a different set of outcomes. Whatever one's views of these proposals, they have a directness of approach not characteristic of all ethical discussion.[14]

To translate even utilitarian approaches into principles of normative behaviour is problematical. It is not simple to find a way of dealing with something as diverse as our own individual behaviour today (shall I go outside in the rain to the compost heap with the potato peelings or put them in with the wastes that go to the municipal tip?) all the way to the whole of humankind tomorrow (how many of them will there be, ought there to be, and to what quantity of resources should each person have access?). Much current action seems to be based on the cost-benefit ratio as an instrument.[15] As we saw in Chapter 3 this is an imperfect technique and says, for example, very little about the distribution of the happiness and good which may be achieved; it also says nothing about any future that cannot be programmed in terms of discount rates. Yet such is the predominance of the western world-view that it has eclipsed most other value systems as a way of re-ordering the world. Students of ethics, however, can at least point to other choices that could be made, both by individuals and more especially by societies. It should be possible to bias decisions against arbitrary choices based on random or temporary factors or whims of powerful individuals; to bias decision-making towards those humans and non-humans who are especially vulnerable to change; to decide always in favour of the sustainable benefit rather than the one-off haul; and always to move against causing harm as distinct from merely foregoing benefits. Some of these principles have found their way into legislation and more into comprehensive reports designed to change world-views (like the World Conservation Strategy, the Brundtland Report and the IUCN/UNEP/WWF 1991 volume *Caring for the Earth*) but it cannot be said that they now outshine their forebears of the here-and-now, business as usual, school.

The non-human world

Although in our anthropocentric way we calmly categorise the rest of the planet as the non-human world, this does not mean that we are released from concern about it. In general, though, there has been a hierarchy of attention based on the degree of similarity between ourselves and the other components of the system: other mammals get the most intensive treatment, then other animals, and thereafter plants, the soil and inanimate things. Of late, the whole biosphere in a functional sense has also commanded the regard of writers on ethics.[16]

Our knowledge of the nature of animals is still accumulating but the more we have, the more it seems true that there are more continuities of biology and behaviour than have in general been recognised.[17] The recognition of an evolutionary continuum between humans and other species seems

123

fundamental to the kinds of judgements we are apt to make about other species of animals. This was not always so: in the West there has been a long tradition of regarding animals as outside the moral universe.[18] Some of this, in e.g. the seventeenth and eighteenth centuries, was largely verbal as philosophers tried to refine the terms of debate, so that their refusal to grant moral standing to animals in the pages of their books was somewhat offset by their love of their dogs or their care over replacing caterpillars on trees. Other parts of it were more practical: St Augustine took over the Stoic tradition of refusing to grant animals any moral consideration and this Christian tradition was kept up by, for example, Pope Pius IX (pontificate 1846–1878) who refused, on those identical grounds, to allow the setting up of a Vatican branch of the Society for the Prevention of Cruelty to Animals. Away from such centres of sensitivity, European colonists killed the native humans and the native fauna with equal facility when they felt like it, and many do not now shrink from the rapid dispatch of spiders in the bath-tub although we may prefer to have lambs made into chops somewhere well out of sight, sound and smell.

One of the turning points in the development of a new sensitivity was epitomised by Jeremy Bentham (1748–1832) who pointed out the essential contiguity of humans and other animals when he argued that the question was not 'can they reason?', nor 'can they talk?' (neither of which can be said of human babies), but 'can they suffer?'. Within that framework, lower animals were held not to be able to suffer, however. In industrialising countries, the social reforms of the late nineteenth century usually included animals, either by prohibiting cruelty or trying to protect wild creatures, or both. The reasons for this greater sensitivity to the fate of animals have been elaborated by many writers and no one argument seems to be pre-eminent.

First of all, there are the feelings experienced by humans for animals. These need no elaboration except to say that they are easily dismissed by the severely rational as being 'mere emotion'. But as Mary Midgley argues so cogently,[19] they are a necessary part of any moral universe, though not sufficient in themselves as the basis of an ethical code. They are, of course, likely to be socially and culturally relative but that does not invalidate the feelings of those who have them. But even in societies with highly developed feelings towards dogs,[20] for example, the use of experimental animals to test cosmetics is still allowed. Moving towards a more objective approach,[21] there is the value (potential if not actual) to us of a species as a resource: for food perhaps or like the nine-banded armadillo which is the only other animal that can catch leprosy and therefore is a test-bed for treatments. And at a slightly further distance towards intellectual and scientific argument, there is the value of biological diversity as material for evolution; this seems still to be important in the age of genetic engineering.

But beyond these ideas which stem from human-centred concerns (which are sometimes labelled 'subjective values' or 'instrumentalist values'[22]) is the proposition that animals have a good all of their own which is completely external to human purposes, i.e. they have intrinsic value. In most people's reckoning this gives them moral standing but not, it appears, equal moral significance in case of conflict. Nevertheless there are those who argue for the equality of all species, whereas others will say that there is a difference between sentient beings and non-sentient ones, with a line being drawn somewhere above the bacteria and viruses. The discussion is carried further by the protagonists of animal rights.[23] They aver that animals have every right to as much moral consideration as have humans and that such standing should be encapsulated in law to the same extent as human rights are thus (somewhat variably) enshrined. Opponents of that view rest their case on the impossibility of animals having interests in the philosophical sense and on their being unable to fulfil the reciprocal obligations which are an essential part of the granting of rights. Instrumentally minded writers are worried that full-scale granting of intrinsic rights to animals would make it impossible for humans to go on living in anything like the ways to which we have become accustomed: we cannot all become Jainists, it is supposed.[24]

Many of the animal-related arguments also apply to other parts of the biosphere and some even to the atmosphere and the rest of the cosmos as well. Plants are the obvious next step, and the larger ones such as trees attract most attention,[25] performing a function analogous to mammals in the zoological realm. Beyond them is the question as to whether the biosphere as a functional whole has a moral standing. Those in favour point to the interconnectedness of everything: without it, they say,[26] humans would not exist let alone have the energy to argue about the future of the Indian Tiger. So there is no real barrier between an individual and the rest of the cosmos and even less so between us and say the plants of this planet.[27] Those against point once more to the ideas of interest and obligation which are inherent in the concept of rights and standing and which the biosphere cannot possess, being non-sentient. By extension, also, not every relationship of interdependence also carries with it a moral bond. Nevertheless our consequent behaviour might have to go no further than Immanuel Kant (1724–1804) who said that we should act as if our maxims had to serve at the same time as a universal law for all the entities that make up the world. 'Think globally, act locally' is today's Green version of the same thing.

Current western ethical systems

We turn now to comprehensive systems of normative behaviour, which lay down principles for the treatment of the environment in its totality. Some systems are extensions of those which deal with people or animals; others are especially formulated in the light of our knowledge of the holistic nature of our environment and our place within it. We consider first those which are ecology-based. These have grown out of the findings of ecology as a

125

science but are now transscientific in nature, having added values and moral imperatives to the original science. Second, we look at those which are theology-based, which in western terms means mainly Judaism and Christianity. Then there is a short section on ethics which derive from radical examinations of our constructions of the world via language, as with Heidegger. Lastly, the question of metaphysics is examined for its relevance to any ethics of the environment.

Aldo Leopold was an academic zoologist with deep roots in the rural landscapes of the USA. He became convinced as early as the 1930s that the emerging science of ecology showed ways of relating to nature that would avoid disasters like those of the Dust Bowl. Leopold argued for the development of an 'ecological conscience', to be elaborated into a 'land ethic' that understood the basic nature of the biosphere.[28] The ethic rests on the principle that an individual organism (humans included) is a member of a community of interdependent parts, with no rights to opt out. The values of that community reside in such features as diversity, resilience after change and connectivity. For Leopold, a process was right when it tended to preserve the integrity, stability and beauty of the biotic community, and contemporary land economics did no such thing, for land, like Odysseus' slave girls,[29] was still property. More recent commentators have pointed out some difficulties with the land ethic idea.[30] At the empirical level, it is not clear just how the manipulative effects of mankind are to be accommodated, since some of them may be stable and even beautiful but have unhappy social consequences.[31] At the philosophical level, professionals of that art have pointed out that the presence of a community fails to generate obligations *ipso facto*. There must be common interests among the members plus a recognition of their mutual obligations for them to be imposed. Further, it can be argued that it is not right to extend ecological concepts like stability, homeostasis and equilibrium to the realm of ethics without proper analysis and qualification. It is certainly the case that these concepts are subject to continual refinement and sometimes radical change. Yet, it is counter-argued, such concepts might provide in some way as yet unspecified a set of objective and cross-cultural norms for the moral assessment of human impact on the environment;[32] further, the nature of the biosphere may be such that, for example, humans and bacteria do have a common interest although they may not be able to communicate this in writing.[33] Although ethical diversity and plurality in themselves may be a moral good,[34] it is difficult to avoid the problems of variability and language. As Aristotle first said, ethics and politics deal with continuous variables and so there could be no certainties in the field of normative decision; similarly we ought perhaps to acknowledge that ecology is not likely to provide the same kind of quantitative and predictive help as the laws of physics and chemistry. It is perhaps always going to be better as a component of attitude formation, but even here there may be the need to formulate different languages and terminologies for ecology as one of the instrumental sciences of human-directed environmental manipulation and as an agent and motivator of environmental protection and preservation.[35]

Beyond this relatively obvious outgrowth of ecological science, another ethic has been put forward, based this time on the convergence of the Gaia hypothesis and the ideas of self-realisation which the West discovered after about 1965. A labelling phrase might be something like 'secular transcendent holism', but plain 'Holism' is less of a mouthful. We recall that the Gaia hypothesis is based on the existence of a number of planetary feedback mechanisms which tend to optimise the conditions for life, though not necessarily for human life-styles, and that they appear to form a genuinely single system. Thus the single term 'Gaia' can be used and the pronoun 'she' is often a corollary, as is the postulate that she behaves in some ways like a single organism.[36] Philosophers have tried then to explain the peculiar features of the human presence within the Gaian system. On the one hand humans may possibly form the nervous system of this 'organism', able to communicate with all of the parts as well as with each other. The flow of information between some sectors and the humans may well be in the form of intuitive knowledge rather than scientific knowledge since we may not yet know explicitly all the ways in which Gaia communicates with her parts.[37] On the other hand, alas, humans might be more akin to cancer cells, proliferating exponentially and 'eating' everything in sight. In that case, modified behaviour propelled by a holistic ethic in which we are 'greened' by Gaian forces is the only route to human survival.

The core of the new environmental behaviour then becomes an awareness of self in which we no longer stop at the boundary of our skins nor indeed perhaps at the limit of our tentacular reach for resources. Instead we are to see ourselves as united with the rest of the universe in a ground of being. One analogy would be that of a drop of water from an ocean: each drop is individual and unique but all are of the same essence as the ocean. This type of thinking has been carried forward by the physicist D. Bohm who uses as analogy the laser hologram in which every portion of the image carries the information needed for the whole.[38] He talks of the material world as being the explicate manifestation of an implicate order in which everything (including human consciousness) is enfolded in everything else. The non-duality of humans and environment thus suggested is reminiscent of many of the religious and philosophical systems of the East.[39] A time dimension may be important as well, for this seems in the western tradition to be unidirectional and thus makes possible the theory of evolution. Secular holists have taken over the concepts of Teilhard de Chardin (which are of course religious: he was a Jesuit[40]) in which there is a progressive infolding of all nature, transforming itself towards some final omega-point of convergence of the consciousnesses of everything. In secular versions, mankind becomes a director of the course of evolution (consider genetic manipulation for example) and thus has special responsibilities. For Henryk Skolimowsky, for instance,[41] we must become the equivalent of priests superintending the unfolding of a sacred drama.

To look for simple rules and cohesive patterns of discussion in the literature and events of ecology-based ethics is very difficult. Perhaps there is throughout an emphasis on process as distinct from objects, in the sense that what we call things are no more than isolated glimpses of something in the process of becoming, just as the bright star is dependent for its luminosity on the darkness of space or just as life holds within itself the promise of death.[42] If you pick up a piano accordion from the floor by one handle, then eventually the other one will follow, though not necessarily directly and not without the odd wheeze. But it's all the one accordion. The human role is seen by some to be determined by Gaian imperatives in which by some metanoic process we shall all change our behaviour; others prefer a continuation of our Promethean traditions, in which we must assume that we are the governors and the innovators but having like all rulers a special responsibility for those whom we rule. Harnessing biotechnology and all other forms of technology, the inheritors of the mantles of Chardin and Buckminster Fuller[43] are anthropocentric to the point of wanting humans consciously to manage the evolutionary processes of the planet: humans act as co-pilots of Spaceship Earth, making management decisions based on information technology. Although starting out from similar bases to the ecological ethics programmes described above, and responding to similar initial environmental pathologies,[44] the holists of this kind are a long way from ecocentric, as distinct from anthropocentric, behaviour.

Theology-based ethical systems

Common to all religions is the idea of a first and ultimate cause, usually expressed verbally as God (or Gods) or the One, or a variant of these words. In many societies, the gods have been identified as being present within all or some of the phenomena of nature and hence as much part of the environment as the air: **pantheism** of this kind, for example, was characteristic as much of ancient Greece as it is of some aboriginal North Americans today.[45] In the West, however, **monotheism** has become dominant[46] and this has been exported along with the other components of the western world-view; we shall here examine the western traditions first and then look at the contribution of other parts of the world.

In the West, early developments about which we know certainly included nature and her processes as part of the focus for worship and ritual and indeed the mystery of the life force was located within such an ecology. The eclipse of these religions by Judaism and then by Christianity, however, removed the mystery to the one God (a trend evident even in Greek times in the dominance of Zeus) who was spatially much more remote than His many predecessors, though knowledge of Him could now be passed down in written form. At any rate, it could be deduced that there was something of a gap between God and mankind and that the close identification of humans with the land was to some extent sundered: 'The land belongs to me, and you are strangers and guests' (Leviticus 25:53). Even the concept of time became different in post-Judaic western religion, for it could not be renewed annually but was linear and each instant was unique. Thus the past could be romanticised as it passed further away and the notion of a Golden Age was born.[47]

The burgeoning of interest in the environment from the 1960s provoked a surge of examination of the Christian position: was mankind indeed alienated from 'the land' for one reason or another, or were we all still part of a continuing Creation which was good, to put it in a highly simplistic form?[48] The first tradition is perhaps the easiest to identify and describe. It derives from the notion that mankind is made, uniquely in the omneity, in the image of God and therefore has the right to behave in a god-like manner towards the rest of the cosmos. This at first sight appears to be the message of the much quoted passages in Genesis I 26–29, where being fruitful, multiplying, having dominion and subduing the earth are the direct commands of God, though not, we must reluctantly presume, in English.[49] This passage was used by Lynn White, a North American historian,[50] as the basis for saying that the 'ecologic crisis' could be laid at the door of the Judaeo-Christian religious heritage of the West, since this passage clearly gave a licence to exploit plants, animals and even every creeping thing. A kind of confirmatory evidence of this view comes in Pope John Paul II's Third Encyclical *Laborem Exercens* in which the forcing of nature to productivity for human ends is seen as a kind of quantitative measure of human grandeur.[51]

A second long-standing tradition is that humans are part of God's Creation just like the rocks and the trees and that no one part of this is inherently superior to another: there is a basic spiritual equality. In this view, both man and nature become co-creators of the cosmos (*cosmos*, it will be remembered, is a world with order) and God is, has been, and will be present in all things. This doctrine of immanence is more sharply focused by the life of Christ, which confirmed that the universe is within God (i.e. pan-en-theism).[52] The rather less abstract symbol of this strand of belief is generally taken to be Francis of Assisi talking of Brother Sun and Sister Moon[53] and preaching to the birds (did he listen as well?); here in Northumbria we have our own ikon, that of the ascetic St Cuthbert being kept warm by Eider Ducks (still known regionally as Cuddy Ducks[54]) after one of his spells of fasting and immersion in the North Sea, whose temperature would have been a severe test for St Francis. Recent interest in this tradition has produced for us figures like Hildegard of Bingen (1098–1179) who celebrates the inherent divinity and beauty of all creation.

This is coupled with warnings about the sins of indifference and injustice to nature, for creation demands justice.[55] She used the term *viriditas* ('green truth') and wrote some prescient poetry:

> Now in the people
> that were meant to be green . . .
> The winds are burdened
> by the utterly awful stink of evil, . . .
> Sometimes this layer of air
> is full,
> full of a fog that is the source of many destructive and barren creatures
> that destroy and damage the earth
> rendering it incapable
> of sustaining humanity.

Much Christian theology is, however, dominated by the concept of the Fall. Any human act is therefore imperfect (and at best provisional) and its redemption is by Grace and probably not in our time.[56] Notably, Francis Bacon thought that a major purpose of the growing science of the seventeenth century was to help regain a pre-lapsarian state. But the overall notion of Sin hangs over all humanity for all earthly time in this view. Since the Bible is the source of this world-view, it can also be seen as the only source of ideas about adapting to it. But faith in the literal truth of the Bible as a source-book for ethics as well as theology is variable.[57]

It seems as if there are two distinct ethical strands which can be woven out of history and dogma, which are representative of current Christian thought about normative behaviour in the environmental context. They relate to the historical traditions discussed above, though with added elements in each. From the first strand comes the common-sense exhortation to recognise the superiority of mankind as being at the apex of creation (so far) but to use the power thus granted with an acute sense of responsibility. This is particularly a Benedictine trait and the example of reclamation of waste places by their medieval abbeys is often cited. So the notion of stewardship is paramount: we are in the position of temporary holders only of the office of steward or vice-regent or overseer and we are required by the Landlord to leave the estate in at least as good a condition as we found it.[58] One trouble here is that the instructions for doing so are nowhere near as explicit as those found say beside the bath in a cheap hotel: how do religious people decide whether it is right to drain swamps or to preserve them for their wildlife?

In some contrast, the Franciscan view has been much amplified by being caught up in the kind of evolutionary mysticism propounded by Teilhard de Chardin. He saw cosmic history as an evolution of consciousness which would end with a total enfolding of the Universe at an omega-point, a final unity with the glorified Christ as Pantocrator.[59] So today's Franciscanism has a much less practical outlook than the stewardship camp (though it is presumably not incompatible with it) in the sense that it is more contemplative and seeks to 'green' (to borrow a phrase) individuals rather than produce institutional change in an overt manner. In this it finds common cause with many non-Christian but avowedly spiritual groups and it is no surprise that a commemoration of the European Year of the Environment in 1987 in Assisi had participants from all the major spiritual faiths of the world. Essentially, this strand of belief plays down the fallen side of humanity and prefers to be celebratory so as to revel in the diversity of all forms of life and the richness of human culture.[60]

The ethical implications of the kinds of beliefs outlined above are not easy to discover, for Christians seem to be able to discover a whole range of proper responses to them: some justify rapid use of resources to create wealth on the grounds that if the Samaritan had not been wealthy he would not have been able to help, whereas others argue for vegetarianism and an extra sweater. There seems to be some concentration, nevertheless, on the preservation of the wild and its non-human inhabitants, on our responsibility to future generations, on respecting the carrying capacity of our surroundings, on the satisfaction of genuine need rather than the inflated demands of consumerism,[61] on the use of appropriate technology rather than everything that the inventors can come up with and sell, and with the need to re-sacralise nature.[62] This last involves putting some of the reverence for life and its mysteries and connectivities back into nature herself rather than allowing it to reside in a remote judgemental sky-god. The poet Gary Snyder phrased it in a rather extreme but cogent way when he said that our [ecological] troubles began with the invention of male deities located off the planet. No wonder, then, that a mystical version of Gaia is attractive to those on the fringes of western religions. Such developments have persuaded radical-thinking but tradition-rooted theologians like J. B. Cobb to develop postmodern religious views which combine the insights of the natural sciences with those of creation-based western theology.[63]

Non-western religions

In the years of high public concern with environmental matters that ended with the UN Conference in Stockholm in 1972, there was much interest in eastern philosophies and religions, and in North America in the beliefs of the native peoples. A contrast can be drawn, for example, between the instrumental view of nature espoused by Anglo-Americans, in which the land and waters are simply resources, and that of the Indians.[64] For the latter, their traditional cultures held that they occupied a sacred space and that all their actions therefore needed sanction from a god or gods, often accompanied by the appropriate ritual. Many expressions of this are found from the late nineteenth-century period of maximum conflict between the

131

two types of culture and with the determined attempts either to marginalise or assimilate the native population, they rather sank from view. With renewed self-consciousness, however, these beliefs are undergoing a renaissance among the Indians themselves and they are being held up by some in the Euro-American community as examples for the nation to follow.

The religions of the North American aborigines (like those of Australia) have never shown much capacity for exportability, whereas those of south and east Asia have always had some fascination for westerners. Thus again in the 1960s and 1970s, Hinduism and Buddhism became much better known in the West and especially for the environmental attitudes which they potentially engendered. (Buddhism will be treated here as a religion since it seems to function as such, though *sensu stricto* it is atheistic.) In Hindu cultures, there is a long tradition of environmental protection, couched under the concept of non-injury or *ahimsa*.[65] In fact, the adoption of vegetarianism and a simple life-style as advocated by Mahatma Gandhi constitutes in itself a predisposition to a relatively low environmental impact; think of the impression on the land that would be made by the Hindu population of India if they all had cars and ate meat.

For Buddhists, the environment is not different from most other phenomena: it can be an object of human attachment and therefore of suffering. Thus an attachment to worldly things that derive from it will end in unhappiness and the law or *Dharma* will ensure that the soul will not escape from the cycle of continual rebirth. There is then a *de facto* ethic of low impact which once again finds expression in an aversion to the taking of life and hence to vegetarian eating. At some stage in its eastward spread from India, Buddhism took aboard many of the essentials of the native Chinese Taoism and the result is known by its Japanese name, Zen.[66] The Tao stressed a quietistic attitude to life: harmony with the cosmos was to be sought by finding its ways and rhythms and adapting to them, rather than striving to alter things and other people. The contribution of Zen has been in stressing the unity of all things and in the primacy of experiential knowledge rather than objective rationality: enlightenment comes suddenly and inconsequentially even though for the strict monk an ascetic regime is thought to be helpful as a precondition for *satori*. Buddhism has combined with native Japanese animist religion (*shinto*) to produce one of the most nature-conscious and delicate aesthetics ever. This too is underlain by a non-dualist philosophy in which the subject-object division of western positivism is absent. This is often summarised in the Japanese phrase '*mono no aware*' ('sensitivity to things'). Emotion is the basis for an awareness of other species, light, weather and eventually of the environment as a whole. There is no vestige of a hierarchy of existence.[67] Since the nineteenth century this has not prevented western values from predominating (indeed, it may have encouraged them since change is always to be expected) although there is now renewed interest in traditional Japanese values and ways.[68] In a broader sense, a progression of concentration upon visual images and their associated emotions can produce the metaphor of nature as a mandala. We might compare this with the well-known image of Earth from space. Such a view of interconnectedness is more explicitly delineated in a central image of Hua-yen Buddhism, the jewel net of Indra. A net is hung which stretches out infinitely in all directions. In each 'eye' of the net is hung a single jewel in whose polished surfaces is reflected all the other jewels, infinite in number. The relationship is one of simultaneous mutual identity and mutual intercausality.[69] There is no centre: certainly not humanity and nor indeed a God: just Being in which the autonomy of individual crystals of being is subsumed.

Islam is monotheistic and based on a book like Judaism and Christianity, and the book (the Holy Qur'ān) is quite explicit in setting humans as stewards of the gifts of Allah.[70] All human activities must be based on the idea that the Earth is only a temporary home (even though man is a superior being) and that to find favour in the next world, our actions must be properly administered as a manifestation of faith. These include justice and piety plus the appropriate knowledge and understanding of environmental problems. The discovery of oil and the more liberal interpretations of some Islamic groups have allowed western world-views to predominate in some parts of the Muslim heartland, but the upsurge of Shia fundamentalism and the imposition of Islamic law may in time bring about environmental changes.

It has to be said that in both East and West many religious traditions have collaborated with human behaviour that is destructive of species and habitat, and with non-sustainable development. In the West, obviously, there has been little sieving of technology and much talk of the conquest of nature; in the East no guidelines have been elaborated for alternative forms of economic and social growth that are ecologically sustainable.[71] In all, some reconstruction of the historic faiths seems to be needed if they are to contribute to an evolutionary *modus vivendi*. It seems unlikely at the moment that, outside the areas of revolutionary Islam, religion as such will play a large part in directly developing normative behaviour, though it may well contribute to the formation of new public ethics of an environmentally related character.

Deep ecology

It is obvious that both ecological ethics and spiritually inspired holism require a change of world-view. A harmony with nature, the avoidance of pollution, the discussion of the possibility of all life having its own intrinsic value, self-realisation rather than economic growth and consumerism, appropriate technology, recycling and thrift, and the organisation of human communities on a regional basis, with great attention paid to minorities, are all found at one point or another in the literature of advocacy. Some, however, have seen this as reformist rather than radical and hence an insufficient response to today's problems. They argue that many of these measures accept the dominant paradigm of humanist instrumentality over nature and are only concerned with tidying up at the edges and avoiding the worst visible excesses. A more radical position is called Deep Ecology and is largely associated with the Norwegian philosopher Arne Naess, who in the 1930s worked with the Vienna School of positivists but who has moved rather far from them.[72] Naess's concept of Deep Ecology collects together the findings of ecological science, the pantheism and process metaphysics of Baruch Spinoza (1632–77), and the historical linguistics of Heidegger.[73] Like some western and many eastern philosophies, Naess constructs a world-view with no ontological divide in the field of existence: there can be, for example, no dichotomy of reality (or value) between the human and the non-human. Similarly, people are knots in a total field and the realisation of Self must not lead to self-centredness but rather to a connectivity with all things which goes beyond mere altruism. This world-view translates into two fundamental norms. The first of these is shared with some of the New Age advocates in the primacy accorded to self-realisation. In this, we must achieve identification with the non-human world: we must learn to 'think like a mountain' and hence let all things be themselves. To harm nature is to harm ourselves. The second norm, also shared to some extent by the previous systems, is that of biocentric equality. The world is no longer our oyster, we share it with the oysters (Table 5.1).

In such a world all things are able to achieve their own self-realisations and thus the space occupied by any 'thing' (ourselves and our technology

Table 5.1 A platform for deep ecology

1	The value of non-human life is independent of the usefulness of the non-human world as resources.
2	The diversity of life forms has a value in itself and humans may reduce this variety only to satisfy vital needs.
3	The flourishing of non-human life requires a diminution of the size of the human population.
4	The increasing manipulation of the non-human world must be reversed by the adoption of different economic, technological and ideological structures.
5	The aim of such changes would be a greater experience of the connectedness of all things and an enhancement of the quality of life rather than an attachment to material standards of living.
6	Those who agree with this have an obligation to join in the attempt to bring about the necessary changes.

Source: Adapted from A. Naess, 'Deep ecology and ultimate premises', *The Ecologist* **18** (4/5) 1988, 128–31.

especially) must be limited to allow all the other things to flourish. One of the great differences between Deep Ecology and the other holisms, however, is its insistence on the value of the experiential as well as the rational, believing as it does that Cartesian dualism is at the heart of most unsustainable relationships within the biosphere. Naess finally collects all his ideas into what he calls *ecosophy*, 'eco-wisdom'. But as his book sets out, he can only talk of 'an ecosophy' because this is a personal system yet one which recognises that many different yet mutually acceptable interpretations of nature are both possible and acceptable. Criticism has been quite strong.[74] There are the obvious questions of the 'how do we get there from here' type, but also a fear that any challenge to the absolute reality of the discrete human individual will lead to some form of totalitarian nightmare: ecological fascism is the label sometimes applied. The counter-argument centres round the opposite view that the glorification of the rights of the individual has in practice led as much to totalitarian societies as those based on notional equality.

The scope for developing Deep Ecology seems quite wide. Recently, other currents seem to have got merged with it: examples are systems thinking, bioregionalism, holistic medicine and healing, feminism and the nuclear disarmament movement. Green politics in its more radical forms is also a likely component. These ideas are likely to share with Deep Ecology at least some of its ethical thinking, if not necessarily its ontological basis.

Towards a radical reconstruction

Many of the commentators on philosophy and ethics remark on the problems of all kinds caused by the almost overwhelming representation of anthropocentrism in western thought and world-view. Since these features of western lifestyle dominate the world in practice, they must be addressed if they are in fact the source of environmental problems. As we have seen, some thinkers try to increase our sense of responsibility, others would go in for mutual coercion, yet others would extend equal moral and legal standing to non-human objects which is in itself logically an anthropocentric act. So there is room for an altogether different way of looking at the difficulties, always bearing in mind that there will be problems of language if we wish to formulate radically novel concepts.

The philosopher most often cited as providing the beginning of such a construction is Martin Heidegger (1889–1976), whom we have mentioned before (p. 80). He attempted to provide a new understanding of what things are and how humans should behave in the knowledge of that understanding.[75] He did not, however, try to formulate a developed ethic, but set an agenda for an all-encompassing *ethos*. For Heidegger, a central concept was that of Being: an event in which an entity could reveal or manifest itself as it really is. All things manifest themselves to each other (as the sun shines on flowers, for instance) but humans have the special capability of noticing that such presences occur. We are actually aware (in a way we suppose beetles and rocks are not) that these entities have a being and also that they might not have one. What then is the authentically human way to live in the presence of all these other beings? For Heidegger, human history and existence constitute a spatial and temporal clearing in which Beings can manifest themselves and be what they truly are, irrespective of their usefulness to us. But being ourselves Beings, we have an essential relatedness to all other beings and therefore to diminish their being is always to diminish ourselves. So here we are beyond the idea of the extension of rights to other components of the biosphere: Heidegger put forward the idea that the core of the relationship was care (*Sorge*) with humans as shepherds of Being, where that Being was a totality of earth and sky, gods and mortals assembled together. He used the example of a jug: it draws together the earth (in the form of wine), the sky (which ripened the grapes), the gods (to whom a libation is poured) and mortals who enjoy the wine and are aware of it all. All these ways of being are significant and no one determines the nature of the others.[76] In other words, we allow ourselves freedom to Be what we truly are when we understand rightly what our place is in the universe, and that is certainly not a position which regards all other beings as a 'standing reserve' of materials.

> Mortals dwell in that they save the earth. . . . Saving does not only snatch something from a danger. To save really means to set something *free* into its own essence. To save the earth is more than to exploit it or even wear it out.[77]

The writings of Heidegger are difficult, both in the original and in translation. There is something of an interpretive industry and there are non-believers who think, for example, that Heidegger is too closely tied to western traditions of thought even though he is perhaps trying to turn around the Cartesian process in which the object of philosophy became knowing rather than living. The opacity of language, though not an unfamiliar thing to academic philosophers, is also clearly a hindrance to acceptability by some. In the end, the message seems to be that in the West especially we must be more open to the possible and that may well mean accepting that there are limits to the sort of rationality to which Aristotle and Descartes have accustomed us.

The end of ethics

The study of normative behaviour looks inwards and outwards at the same time. In the case of the former, there are two especially popular windows. There are those who say that basically the human concepts of utility and justice as elaborated in the West are all that is needed for a viable and valid environmental ethic. But a problem here is the fragmentation of advanced societies into systems such as law, education, economy and religion. The need for an ethic produces a level of debate in each. But since no one function system equals the whole of society, the level of resonance in any one function system does not necessarily produce a valid ethic for all. Thus others argue that some new metaphysical insights (in particular going beyond the present range afforded by the various brands of humanism) are needed.[78] In the latter field, the non-separation of everything which is one of the more startling results of modern quantum theory at the particulate scale is a possible starting point for the discovery of intrinsic value in non-human entities. Here, if the self is valuable, then all else is equally valuable.[79] This argument can be extended to suggest that the universe in its entirety possesses a measure of self-hood in being a self-realising system. It does not have a purpose or *telos*, but it is dynamic and unfolding just like smaller scale manifestations such as an organism. This idea of self-realisation can be extended to inorganic things if we include the system in which they are embedded. Humans can add an extra dimension since we alone can understand our relationship with greater wholes as well as smaller parts.[80]

It is difficult to see a discussion of the Copenhagen Interpretation of quantum theory being the basis for a Greenpeace call for funds.[81] But the movement towards the development of a better public ethic brings in various of the ideas discussed in the last few pages. They are neatly put together by Charlene Spretnak at the end of her book on spirituality in Green politics[82] and they act as a good overall focus precisely because they bear no very clear relationship to what she says earlier in the book, i.e. they are as valid in a secular context as in a transcendental one. She calls for ecological wisdom, grassroots democracy, personal responsibility over lifestyle, non-violence, community-based economies, post-patriarchal values, respect for diversity, a global responsibility and a vision for the future which focuses on the quality of life. Although there is a humanistic bias in these recommendations (which may be compared with Georgescu-Roegen's bio-economic programme, see note 50, p. 175), they might well be a good start along even the most radical of non-anthropocentric roads towards an altogether different basis for ethics.

ENVIRONMENTAL LAW AND POLICY

Once an ethic becomes public then most societies want to incorporate all or part of it into a formal system of policy and law. Some societies may choose to emplace parts of the ethic by custom, relying on the general pressure of society to enforce a practice; so the law may be unwritten but no less powerful for that. Here, however, we shall be mainly concerned with the written legal systems of industrial societies and further with the problems of international law. In particular, we shall need to look at whether the apparatus of policy-making and the law can deal with the complexities of the environmental revealed by science and also fulfil the aspirations of many people for access to resources and for a healthy set of surroundings.

Principles

The formation of law and policy can be broken down into three main distinct categories of activity. There is first **jurisprudence**, which is the empirical response of society to its regulatory needs, in terms of social and philosophical principles, data collection and interpretation, and methodologies for processing legal activities. From this angle, therefore, the law itself is an applied social science and must needs interact with the other social and natural sciences. Second, there is **policy-making**. Policy is the expression of the aims and principles of a society (which may be at various levels of spatial scale and generality) and incorporates the outcomes of past decisions into its feedback mechanisms. Third, there is the **positive law**[83] which encompasses the legal provisions themselves (such as rights and duties), the procedural rules for resolving conflicts and organisational rules for the conduct of public power.[84]

Within this very general framework, there may be considerable differences of approach. There is, for instance, a great divide between those legal systems which are based on English common law and those deriving from Roman law. In the former, a precedent-based *modus operandi* is adopted which proceeds from case to case in a pragmatic rather than an intellectually satisfying way. There is also a tendency to decentralise under such a system. The other approach, much favoured in continental Europe, aims for certainty as a fundamental goal: the law is seen as an antecedent inerrancy which keeps ahead of the facts. Here, there is no room for interpretations of the law by those who administer it.[85]

One particular influence on the development of positive environmental law has been the general nature of technical law. The lawyer expects there to be a given set of rules which he/she can apply, based on a clear socio-philosophical foundation. Thus the evolving nature of concern with the environment in recent years, coupled with the emergence of new concepts like the ecosystem and novel problems like acid rain and carbon dioxide, has taxed the workers in jurisprudence to find a proper and workable response of the kind to which they have been accustomed in, for example, the law of property. They have had in the past tools such as land use controls, easements to preserve scenic attraction and taxation, but few developments to cope with, for example, the necessity for the integrity of the whole nuclear fuel cycle. In fact, over the last 15–20 years, the interests of the human environment have in legal terms been perceived to be largely the protection of basic renewable resources against various forms of pollution and degradation. Where these phenomena have been restricted in scale then the relevant scientific and technological data have usually been relatively easy to gather and quantify; where regional and transnational, even global, contaminants are implicated, then difficulties have been greater.

Even so, the law in many countries has attempted to locate, contain and deal with the diverse problems of, for example, pollution within just one of the media in which it occurs, like land, water or air. In the UK there has been the jurisdiction of the National Rivers Authority, Her Majesty's Inspectorate of Pollution and the duties of local authorities to deal with air quality and with waste disposal. One example of the complexity of legal constraints is the question of controlling mercury discharges into the North Sea. In Britain, this is overseen by international law (the Paris Convention, ratified by the UK in 1978), EC law (which placed limits on discharges by the chlor-alkali electrolysis industry, in a directive of 1982) and UK law, under which the National Rivers Authority must give consent before new discharges are allowed and may impose conditions.[86] Harmonising these different levels of law is always difficult, especially if they spring from different traditions of legal history (see above) and the stage is then set for laggards who claim that the environmental and technical conditions under which they operate are inappropriate for the enforcement of the control measures.[87] Objectively, it has also to be realised that the law may have limits to its effectiveness, as shown by the operation of the Environmental Protection Act of 1969 in the USA.[88]

Transnational law: the EC

As an example, the evolving role of the EC in setting standards among its member states in environmental affairs may be briefly mentioned. Environmental policy started in 1972 and was made fully legal by an amendment to the fundamental treaty in 1985. Title VII of Part 3 of the EC Treaty now espouses:

- the protection and improvement of the quality of the environment;
- the protection of human health; and
- the prudent and rational use of natural resources.

So for the prevention of problems, the EC lays down that Environmental Impact Analysis must be carried out for major projects; that there must be at-source control of pollution and full integration of environmental policy into socio-economic development. Pressing problems are seen as the pollution of the atmosphere and the sea, hazardous chemicals and wastes, the deterioration of urban environments and the protection of the soil. Initial evaluation of the policies and their enforcement by member states suggests that implementation is to say the least spotty and the EC has moved to appoint community inspectors, allow NGOs access to EC institutions to inform of infringements and allow the EC itself to instigate proceedings against transgressors.[89] It is noticeable that the EC has in general adopted the US practice of resolving conflicts in a series of courts (i.e. adversarially) rather than the UK practice of a public inquiry, which is inquisitorial though time-consuming since almost anybody can decide to appear.

Where international matters are concerned, the oceans form a focus for the exercise of legal and diplomatic talents. There is a long history of treaty-making and other arrangements for the management of fisheries and for the protection of wildlife, but since the 1950s marine pollution has attracted the most attention: by the late 1980s there were over 70 multilateral instruments of different types.[90] These fall into three main sectors: vessel-source pollution, land-based marine pollution and dumping. These reflect not only the variety of wastes that the oceans receive but the difficulty of dividing up pollution into environmental sectors since anything is likely to end up anywhere, so to speak. The whole issue was given impetus by the UN Conference on the Law of the Sea which in its 1982 final Convention concentrated on these three sources and their management, though the Convention recognised that there were other origins of contaminants, such as offshore petroleum exploitation, mineral resource development outside national waters and the atmospheric transport of pollutants which then were scavenged out over the seas of the globe. The general agreement, however, is that many of these instruments are less than totally effective in dealing with the problems they were set up to solve. J. E. Carroll reviews examples like the Boundary Waters Treaty between Canada and the USA and the UNEP Mediterranean Action Plan and argues that they both leave enormous gaps in their coverage.[91] In the first case, the problems of acid runoff and the gross pollution of the Great Lakes have been little tackled; in the second, there may be amelioration of gross effluents round the immediate shores, but insidious inquination from further inland is likely to be unchecked. In other words, there is at present a large gap between the direction of environmental diplomacy (towards more agreements) and the pace of accumulation of change: the law is not fast or flexible enough to keep up with economic change. Thus it is possible to argue that reformist attitudes to environment will never save the planet from destruction of some of its vital parts.[92]

But for the law itself there seems to be a fundamental divide. On the one hand there are those who view the law in itself as having an instrumental function of an almost metaphysical type, catering to one of the absolutes of human society in the shape of the need to regulate the behaviour of individuals so that they may not derogate from the greater polity. On the other are those who see the law as part of the socio-political framework dedicated to enabling society efficiently to carry out its chosen goals. In this latter context we find most of the current research into formulating systems of environmental regulation of pollution and resource use and ways of ensuring compliance with regional, national and international policies.[93] To that extent, therefore, although words like 'ecomanagement' may be used of an integrated process of policy formation and legal enactment, the law forms part of the instrumental view of the environment that is characteristic of, though not confined to, the western world-view.

SOME PROBLEMS

At the beginning of this chapter we mentioned that there were problems of scale in formulating enforceable behaviour. None of these seems to have gone away in the last twenty-three pages. Much law after all originated in the ideas of applying constraints to individuals and then aggregating upwards. But as we saw in the discussion of ecology, the whole is often greater than the sum of the parts and legislating for these emergent qualities is clearly difficult. In the last twenty years or so, it has become apparent beyond any doubt that humankind can affect the whole globe. That is to say that we possess the capability of changing the global whole, especially in the field of climate which even non-ecologists recognise as being rather important in all sorts of ways. In terms of regulations, the problems for managing such alterations centre around features like:

- The location of e.g. climatic processes in global 'commons' which nobody owns and therefore nobody has primary responsibility for.
- The difficulty of drawing up and enforcing rules which have to cater for a wide variety of institutions. In the case of the ozone-destroying chlorofluorocarbons (CFCs), some may be made by state-controlled nationalised industries, others by small firms willing to move from country to country to escape regulation.
- The problems inherent in legislating for a future situation which is not precisely known. In the late 1980s to early 1990s, for example, the likely magnitude of the response of climate and sea-level to increased carbon dioxide concentrations in the atmosphere has been carefully modelled, as was the outcome of the formation of 'holes' in the ozone layer due to its dissociation by chlorine from CFCs. Yet few governments showed any signs at that time of taking the problem seriously enough to start to formulate laws restricting the emission of carbon dioxide. Only after 1992 and the Rio Conference were there moves towards greater control.

Another question which arises is that of the purpose of normative behaviour. In other words, what are ethics *for*? Are they simply for the short-term regulation of the supply of resources, the use of nature for processing or storing wastes and the maintenance of pleasant surroundings for the fortunate? Or, is the long-term survival of *Homo sapiens* a desirable purpose to adopt? Or, is this much too anthropocentric, and so the long-term aim should be the recognition of the intrinsic value of everything non-human and so, basically, participation in a cosmic evolutionary process?[94]

Without some view on these questions, there is a default position: we shall look no further than the next stop-light on the street and hope that science and technology will change the light to green. This way of proceeding has a long historical sanction but any examination of the types of knowledge we possess should convince us that it is merely algedonic: the rewards and punishments which are offered are an incomplete explanation of the metasystem in which we operate and ones which cannot respond in a linear fashion to the fact that the traffic might be about to exceed the capacity of the roads.

6

ONLY THE ROAD. . .

There is a thirteenth-century altar-piece in the cathedral at Toledo which bears the inscription:

¡Caminantes! No hay camino, hay que caminar.[1]

Equally here, there is no sense of Conclusion, simply a resting-point at which we may, from a slightly elevated place perhaps, look back at where we have been and also survey the different possible paths onwards.

In Chapters 2–6 we have looked at the more or less conventional approaches to the construction of 'environment' as we in the West (and those who are our cultural followers) understand it. In most sections, it is true, some of the alternatives to the mainstream ideas have also been considered. In this last section, though, I want to address the question of whether in fact all these ideas are much too partial, that in effect they do not go far enough in enhancing our understanding of the relations of humanity and the rest of nature to be useful at either an intellectual or a practical level. In other words, we need to examine some trains of thought which go beyond the very modes of thought and reasoning which have so far been employed. Inchoate strands of such notions have emerged from time to time so far, but here is the place to make them much more explicit. So these will be examined in the main middle section of this chapter before, finally, we return to the over-arching question posed by the natural sciences and their emulators: is the world real or is it totally a construction which we may remake as we wish? Does an environment envelop us all and enforce outer limits to what we may do? Or are such notions constructions of our minds which may legitimately be replaced by others?

The main set of themes around which the central discussion section will be centred are:

• The feminist contention that much of the type of thought discussed so far is ineluctably patriarchal: that most science, literature and the arts, philosophy and certainly religion, is constructed by men and mainly for men.

- The humanist argument, in which the quality of self-knowledge of humans and their communication with each other about the world is central. At one extreme there is the thrust towards the Dark Green position which aims at an emotional identification of humans and the non-human world.
- A more extended examination of one non-western tradition of thought to see if it represents resources for the whole of humanity in any search for a sustainable relationship with the non-human world.
- A glance in the direction of those strands of poststructuralist thought which aim to disestablish some of the tenets of western discourse such as the identity of the self, the possibility of fixed meaning in language and indeed the very idea of there being foundations to knowledge.
- A discussion of the special claims of science to give us the kind of information upon which we can rely in a pragmatic and predictive sense and which is free from all the baggage of culture and cognition that we associate with humanistic constructions of environment.
- A review of the position of humanity in evolution at cosmic and ecological scales and the emergence from it of certain traits of behaviour which seem inimical to the sustainable humanity–nature relationships of which we speak.

FURTHER EDUCATION

In this section we pull out of some of the earlier discussions some more fundamental notions which make us put much existing knowledge in a new light. Some of these ideas take a JCB to the very foundations of what we think we know and so are strongly disputed: not all the turns of the arguments and counter-arguments can be reproduced here.

Feminism

Representation of the world, like the world itself, is the work of men; they describe it from their point of view, which they confuse with the absolute truth.

(Simone de Beauvoir, 1908–86)

This section addresses itself to the idea that our whole way (in the western rationalist tradition at least) of examining our relations with the natural environment is corrupted by the fact that it derives only from the actions and ratiocination of men and not of women. It is no surprise that the feminist movement of the post-1960 period has seized upon the 'environmental crisis' as one of its subjects for analysis and the resulting set of views can be called, by way of a shorthand term, 'ecofeminism'. The term encompasses 'the struggle against the oppression of women and natural environments'[2] and has in common with Murray Bookchin's work (see p. 59) the ideas of control of women by men and of nature by both; women are in the middle of dominance hierarchies.[3] The basic principles of ecofeminism are holism, interdependence, equality and process. Most of these are familiar from other contexts: holism, for example, from ecological science and the Gaia hypothesis, and interdependence from those ethical traditions which accord equal value to the human and non-human. Equality spurns the dualistic and hierarchical societies which are common and suggests that these are the most likely to follow paths leading to the degradation of nature. An emphasis upon process affirms that ends do not justify means and the quality of all relationships is an end in itself.[4]

The links between women and nature are subjected to historical scrutiny by ecofeminist writers in order to see when and where the situation of a white male dominance arose. The obviously patriarchal nature of Judaeo-Christian societies in Europe before the Renaissance compared with what we know about their prehistoric equivalents in the Mediterranean is one major theme. This strand is taken further by Merchant for the period of the scientific revolution,[5] where she shows that as science developed, women were progressively excluded from it. The development after 1500AD of a world-view in which nature was among other things a source of disorder meant that women could be identified with that side of nature and thus had to be controlled, just as nature ought to be brought under man's hand. Women's previous links with the natural world (healing skills, for example) were often then labelled as witchcraft. More recent psychological approaches suggest that female development in childhood is based on identification with others, while that of boys is based on separation from others so that men fear things which are different from themselves. This anxiety leads to a need to exploit and control other people (especially women) and the non-human features of their lifeworlds. In the eyes of some, this is at the basis of the enterprise of science itself: it can be seen as essentially an extension of the male ego: penetrative of the mysteries of (an essentially female) nature and ineluctably destined to lead to control. At its worst this leads to the weapons industry, which is the desire for male domination writ large, not least in the symbolism of weapons.[6]

At present, ecofeminism defies any narrow definition and so the reconstructions of relations proposed are themselves plural. It is argued that women are more aware of the natural world than men and so women should create a culture which honours both women and nature: equal responsibility for childcare might alter the attitudes of boys towards both men and women. At heart, however, ecofeminism seems to be phenomenological, 'fundamentally a feeling experienced by many women that they are somehow intimately connected to and part of the earth'.[7] This outlook surfaces explicitly, however, in only one of the two major versions of ecofeminism which have emerged in the last 20 years.[8] These are:

1 Cultural ecofeminism. This stresses the biological, historical and experiential links between women and nature. An alternative culture would revalue, defend and celebrate what patriarchy has devalued, including the feminine, the non-human, the body and the emotions. A new spiritual relation to nature is also a frequent quest.
2 Social ecofeminism. Here, nature is viewed as a political rather than a natural category and it falls along with many others that are the victims of oppression by male dominance. Gender structures and the dominance of nature have, it is argued, helped to hold different forms of oppression together.

There is, however, a common oppositional project in emplacing a vision of a society beyond militarism, hierarchy and the destruction of nature; an integration of these approaches, suggests K. Warren,[9] would give us a transformative feminism which would centre about such values as:

- the recognition of all forms of oppression, including that of women and of nature in societies;
- a theory which includes a recognition of the diversity of experiences but which looks also for maximum intersubjectivity of felt experience;
- a rejection of the apparent logic of androcentric modes of thought and practice so as to press for a psychological reconstruction of many attitudes and beliefs about ourselves and the world; here the analysis goes beyond Deep Ecology and says that it is not simply anthropocentrism that is at the root of environmental problems but androcentrism;
- giving a central place to values which are lost in today's ethics, notably care, reciprocity and diversity;
- challenging some of the ways of scientific and technological research, in particular those which lead to destruction rather than preservation.

There are some contradictions in current ecofeminism, as might be expected of a set of ideas in progress. Prentice, for example, argues that it is based almost monolithically on the assertion that biology is destiny: that domination and oppression are basic to men and that it has little enough to do with social structures which, being human, can be altered. According to Prentice, ecofeminism is 'the product of North American privilege' and therefore not part of a bottom-up strategy of political change.[10] Such transformations would rank environmental destruction alongside issues of class, gender and racism all of which in her view derive from a capitalism one of whose servants is science. B. Easlea has put this in a broader framework[11] but noting that science celebrates hardness and conflict, he cites Bertrand Russell's view that 'all modern scientific thinking is at bottom power thinking'.

Through this seeking of power, men wish to exert control over each other, over women and over the environment partly because of material conditions and class conflict and partly to demonstrate an otherwise elusive masculinity. So women are here tied into a complex of other factors. In a long historical perspective, W. I. Thompson has reconstructed male–female relationships which appear in the form of myths, concluding that they reflect the growing dominance of the male. This latter ends (at present) in the stockpile of ICBMs and these represent a delusion of power. Only a 'creative destructuring of the . . . industrial civilization that humanity has already outgrown' will, for Thompson, avert the ruination of western civilisation.[12]

It remains to be seen how much progress ecofeminism makes. If other fields are any guide then some changes are likely because of this thinking, but unless the currents join with others then the all-embracing integrative revision envisaged is likely to be something of the far future, if indeed it happens at all. If, however, the feminists are right, nearly all of our previous discussions of this topic, and many others, have been deprived of the potential contribution of most of the women of the world.[13]

Anthropocentrism and ethics

The relations between the formulation of an environmental ethic and the location of value are crucial and complex. Naively, we might argue that since the whole notion of ethics is itself a human creation, there can be no discussion of it which is not *ipso facto* anthropocentric. Yet of course this has not prevented a whole group of people wanting to formulate an ethic on the basis that non-human natural entities have a value independent of human ratings.

In any human-centred view the individual is at the heart of the demand for environmental products, but an assurance of a steady flow of supplies means that the capability of the supplying system is every bit as important. Therefore the emphasis is placed not on the needs of the individual but on the holistic capability of the systems to fulfil the needs of entire societies in the future. The stability which is sought, however, is not the same as some models of ecological 'balance' or 'stability'.[14] Thus environmental impact need not be eschewed if the resulting system is a stable one with a sustained yield;[15] non-renewable resources may be used if steps are taken to provide reasonable substitutes. This may also entail some kind of population policy in order to limit the demands placed upon any system. Overall, the idea of a weak anthropocentrism avoids the problematic area of attributing an intrinsic value to non-human objects.[16] It relates to the cosmologists' 'anthropic principle' beyond which the strong version asserts that the Universe must produce beings capable of observing it. In one sense, then, they are a stage in its evolution and there can be no duality between them and the rest of the Universe. Being contiguous with pre-existing nature then, contiguous value is extended to all that pre-dates and surrounds humanity.[17] The Deep Ecology movement is to some extent a translation of that into a platform and slogans of the type reproduced as Table 5.1,[18] though its critics declare that it is too close to mysticism or that it ignores vital differences between anthropocentrism and androcentrism.[19]

Given the dominance of western thinking and technology,[20] it may seem fruitless to evaluate alternatives but given also the current fact of the 'global village' when one idea could quite quickly sweep to prominence via ubiquitous electronic communications, then perhaps we should mention any other ways of viewing these relationships.

A yen for Zen?

Eastern cultures have long been known in the West for their non-duality and their different view of the relations of cause and effect. The image of the jewel net of Indra (p. 134) is one such. In an environmental context, possibly the most developed notion is the Japanese concept of *fūdosei*. This has been translated into French by A. Berque as *médiance* which he uses as a shorthand term for the concept of the occasion of human existence *in toto*. As is traditional for Japanese culture, he argues for the non-dualism of man and nature: *fūdo* is a relationship which partakes of both the subjective and the objective, so that it is at once the ecological character *and* the symbolical meaning to humans of a given place or *milieu*.[21]

There is therefore an insistence on pure experience which places the relation between a subject and its world as the first reality. Within this, the subject does not bear a definite place according to which everything else has to be ordered: the world is not something of a lower order to be challenged, altered and conquered.[22] Rather, it is a set of relationships which bear the weight of constant adaptation. In his comparison of East and West, Hwa Yol Yung derives the Japanese attitude from the aesthetic and reverential attitude to nature which is characteristic of Zen Buddhism.[23] This in turn is an amalgam of the practicalities of the Chinese mind with the speculative turns of Indian thought: Mahayana Buddhism come to fruition in Chinese soil. In the Japanese version, nature is a being-in-itself rather than an object-for-something-else; the word for nature (*tzu-jan* in Chinese, *shi-zen* in Japanese) can be translated literally as self-thusness. The way of communication of this attitude is via the poetic rather than the practical: sensibility becomes the primary vehicle of embodied consciousness rather than the reified conceptualisations of western post-Cartesian and post-Baconian thought.[24] In one way, this becomes a plea for humanity to re-establish themselves as natural beings, inhabiting the cosmos as participants rather than as observers. A first step for the Zen practitioner is clearly contact with nature, and the provision of adequate areas where nature can flourish thus becomes a starting point for this world-view, and one where it and western instrumentality can intermesh. A second is presumably an investigation of the nature and meaning of experience East and West: a cross-cultural hermeneutics of environment.[25] This aims towards a transparency of communication which is made more difficult by problems of language itself and by the cluttering of any channel of communication with the noise of power relationships.

The problem of language

Perhaps one of the key revolutions in thought in the twentieth century has been in the postulated reversal of the role of language *vis-a-vis* the 'outside world'. Whereas it was once thought that language was a means of attaching verbal labels to objects and abstract entities, there are now many scholars who prefer the idea that language is primary and defines reality: what is 'real' therefore comes to us only through language. Benjamin Whorf (1897–1941) puts it succinctly:

> Language is not simply a reporting device for experience but a defining framework for it.[26]

The key figure in this transformation is the French exponent of structural linguistics, Ferdinand de Saussure (1857–1913) followed by the semiologist Roland Barthes (1915–80). For de Saussure, there was a **signifier** like the word 'apple', and a **signified** like the concept of an apple. The relationship between the signifier and the signified constitutes a linguistic sign and language is made up of these signs. We must note that the sign is not born of necessity; it stands for something only because of use and convention. Further, each signifier get its value only because of a differential place within language as such. Development of these ideas beyond those of its founders has suggested that there is no one-to-one correspondence between linguistic propositions and reality: language again is primary.

At any rate, it is language that creates the possibility of subjective consciousness: thought is the movement of signs. Language becomes the source of meanings and of 'truth'. Only within language can the world (as well as ourselves) be formed as an intelligible reality.[27] But there is always an indeterminacy of meaning because any signifier can always receive retrospective signification: there is a chain of words, with the possibility of slippage along the signifying chain. One potent source of this is the use of metaphor. Although metaphor has traditionally been seen as a rhetorical device, it may function in a much wider context. Since language works by the transference from one kind of reality to another, then it is perhaps metaphorical by its very nature; meaning shifts around, using metaphor as its main process. There is, for example, no limit to the number of metaphors for any given idea; similarly, metaphor states one thing while requiring us to understand another. Many of our basic ways of handling language are metaphorical in ways we no longer question: organisations are spatial, with ups and downs; theories are buildings, with foundations and frameworks;

time is money; leisure is to be filled; we are 'in' or 'out' of work. Many kinds of discourse, therefore, are structured by something which needs considerable interpretation because its meaning is by no means fixed: time does not have to be money, nature might be more like a mobile than a machine, for example, and so on. The major point of such excursions is to understand that in language, the signifier does not yield up a meaning directly, as a mirror shows an image.[28]

The views of indeterminacy of meaning reach something of an extreme in the works of the French postmodernist philosopher Jacques Derrida (1930–), who argues that signifiers and signifieds are continually breaking apart and recombining; further there is a temporal element to the process: earlier meanings are modified by later ones. If we read one poem, then every other poem we read may have its meaning altered by that act. Going back to Saussure, we add the fact that meaning is also present in what a sign is *not*, so that reading is accompanied by a constant flickering of distant lightning as much as it is the counting of sheep jumping a fence.[29] Derrida's way of examining a text is called *deconstruction*. This aims to show that the author's conceptual distinctions on which the text relies are inconsistent and paradoxical. Metaphors in particular are singled out for searching attention. Derrida argues that the search in the West for some sign which will give meaning in an ultimate sense to all other signs (the 'transcendental signifier') is meaningless. Meaning must then be plural since a text is undecidable in its meaning and there is, further, no self prior to or independent of the investigations.[30] Neither Derrida himself, nor any of his more accessible followers and exegetes have looked directly at how such thinking might affect our view of the relations between humanity and environment. He himself, however, has deconstructed Jean-Jacques Rousseau (1712–78) where the topics of nature and culture are under examination. Rousseau starts out with nature as a primitive stage, with simple human societies living happily, who then add cultural complexity. Culture then adds to nature and substitutes for it, but Derrida argues that each time Rousseau uses the term nature he describes it in terms of nature supplemented by culture or indeed in terms of a nature/culture polarity with the first term being a better state than the last. Nature thus must become something which is never un-supplemented. I suspect this may translate for present purposes to something like the notion that 'environment' and 'nature' can never be used as terms to postulate some primitive state of the biosphere: what we can relate about the biosphere of say two million years ago is still as much a reflection of the present use of words and especially of metaphor and of the shifting entities which we call 'self' as of any truth of what the world might have been like two million years ago.[31]

We have then to watch our language. What do the signifiers like 'environment', 'nature' and 'human' actually lead to in terms of signifieds? If we follow some of the French postmoderns, then behind the façade of contemporary usage there is a great deal to be unpicked. For example, no one interpretation can have higher standing ('be privileged above') than another; there is no datum line, for example, against which to make judgements about human impact on nature; our verbal communications are full of gaps, of things which we are not being told; and no one narrative (including that of natural science, presumably), ought to be empowered above another. No wonder that Derrida and his school have been criticised for a lapse from post-Enlightenment humanism to a position which encourages

Table 6.1 Global mean energy flows for various natural and human-induced processes

Process or event	Energy flow (cal/m^{-3}/day
Solar energy to earth	7000
Solar energy absorbed by earth	4900
Primary production by plants	7.8
Hurricanes	4.0
Tides	1.54
Animal respiration	0.65
Cities	0.45
Forest fire	0.3
Fossil fuel	0.11
Urban fire	0.065
War (non-nuclear)	0.05
Floods	0.04
Earthquake	0.001
Volcanoes	0.0005

Source: J. F. Alexander, 'A global systems ecology model', in R. A. Fazzolare and C. B. Smith (eds) *Changing Energy Use Futures*, New York and Oxford: Pergamon Press, 1979, vol. II, 443–56.

Table 6.2 Human appropriation of net primary productivity (1980s)

World NPP	Organic matter in Pg (× 10^{15}g)
Terrestial	132.1
Fresh water	0.8
Marine	91.6
	224.5
NPP used directly by humans:	
Plants, animals eaten, wood used	7.2
NPP used or diverted by humans:	
Croplands and other land conversions	42.6
NPP used, diverted or reduced:	
Used or diverted	42.6
Reduced by conversion	17.5
TOTAL	60.1

Source: J. M. Diamond, 'Human use of world resources', *Nature, Lond,* **328**, 1987, 479–80.

first anarchy and then chaos. Possibly the central lesson of postmodernism is that there is no intellectual tradition that has a privileged authority. Whereas modernism saw culture, science and society as a Grand Hotel (linked enterprises organised to a commanding theme), postmodernism sees them as a shopping mall, an infrastructure which supports disparate enterprises in which appearance is all. So it rejects all 'metadiscourses like science; avers that truth is contingent; and celebrates variety and eclecticism'.[32] Whether it represents a grand change which will have consequences for the kind of story told here is too early to tell. J. Cheney argues for *place* as the scene of a set of locales for truths which are socially negotiated for each setting, and which in turn contribute to our constructions of understandings of self, community and world.[33] One essential ingredient seems to be local ('bioregional') narratives, not as givens from which behavioural injunctions follow but as achievements to be striven for as part of a healthy community.

Extensions of idealism

Tables 6.1 and 6.2 show the extent to which, in scientific terms, the 'natural world' has been assimilated to that of the human. Table 6.1 shows the natural fluxes of energy on the planet as compared with those that are directed by human activity; the second is a sub-set dealing with living matter. This very clearly delineates the fact that of the energy fixed as organic materials, we now appropriate a very substantial portion to our own uses.

Given this finding, and also the further ambitions of mankind for material acquisitions and technological development, all the mental constructions of the environment assume an importance as great as those intersubjectively agreed by natural science. But important for what or whom? We have here to assume that it is a legitimate long-term purpose to ensure the future of life on Earth, including our own species, otherwise there seems no point to the question at all. In this light, we must look back over the materials of Chapters 3–5 and ask if they are adequate approaches to grand-scale questions such as *are* they near the 'truth' or are they too fragmentary to get near 'truth', and, perhaps, would we know 'truth' if we saw it anyway?[34] Approaches to these questions are made by, among others, the 'grand theorists' in the humanities and social sciences and we need to look at what they have to say on these questions.

Theories of the 'human sciences'

It will be a convenience to use this term to encompass both the humanities and the social sciences, as is done by the German term *Geisteswissenschaft* (as opposed to *Naturwissenschaft*), since they can often be treated together. It is fair to say that the writers in this field have generally ignored 'environmental' concerns[35] and we shall have to use very general statements or else adapt their words to the material, with consequent dangers.

Overarching 'grand theories' in the social sciences have a long history: environmental determinism is one such, as at the other pole is the historical materialism of Karl Marx. In the post-1900 years, there have been attempts to carry forward such endeavours, just as there have been statements to the effect that everything is messy and contingent and no such all-encompassing theories can possibly be true. So existentialism, modernism, structuralism have all paraded past succeeding generations of scholars and *literati*, without exerting much effect upon those who chose to write about the relations of humans to the natural world in a positivist framework. But since at any rate some major attempts at theory go to the heart of all human actions we need to see if they have application in the sphere of this book.

But it is necessary to say that the whole notion of the social sciences has been much affected by the progress of the natural sciences and in particular by their success in formulating laws and in making temporal predictions. A major question still remains for many philosophers of the humanities and social sciences as to whether social objects can be studied in essentially the same way as natural ones. In general, for the positivist, science is outside society; for the student of learning, society is outside science. The weaknesses of one find antitheses in the strengths of the other. One approach to the problem has been the definition of a **critical naturalism** which does justice in social science to both approaches. This involves the recognition that laws and similar empirical regularities are impossible outside closed systems and that explanation in open systems can take the following form outside closed systems.[36]

The production of meaning is thus governed by law but not determined, for the conceptual activity of human agents is necessary for the existence of social structures and necessitated by them. A critique is applied to this process in terms of what it includes and what it leaves out. All this is itself a moment in the process it describes, so that nothing is excluded; it is critical and it is necessarily self-reflexive. This all adds up to the idea that the human sciences can be sciences in the same sense as the natural sciences but not exactly in the same way.

The existence of naturalism, however, while strengthening some of the social sciences has not prevented the growth of frames of investigation which are more self-referential in the sense that they ignore the dichotomy between positivism and hermeneutics, claiming that they both rest on an ontology which has been rendered obsolete by new ways of looking and thinking. It is also a theme picked up by Vico, who might be a pivotal figure between pure forms of phenomenology and positivism.

Structure and agency

In the school of social theory called **structuralism** we have an attempt to recover by intellectual processes the ultimate basis from which the whole of the variety of human behaviour is generated. The attempt moves beneath the visible and conscious designs of active human behaviour in order to reveal an essential logic. This logic binds all the secondary designs together in an enduring structure. Another way of thinking about the social structures in which humans are located is that of **structuration**. Proposed by the sociologist A. Giddens, it focuses on the intersections between capable human agents and the wider social structures in which they are necessarily implanted: people make their own history but only under definite circumstances and conditions. Hence, social interaction entails communication, and has to be grounded in structures of power and legitimation; everyday life is bound up with long-term social institutions, and scales of change are contingencies that are present every moment in the evolution of social systems. Thus structuration focuses on space–time relations in a way most social theories do not: it is particularly useful for the study of the dynamics of the social side of human–environment relations but tells us more about the complexity of changing that relationship than about the nature of the relationship itself.[37] The flexibility of time and space is encapsulated in the idea of the *production* of e.g., nature and of space in an economic system. N. Smith, for instance, argues that the uneven development of any part of the world is 'the hallmark of the geography of capitalism'. Smith says that capitalism transforms the shape of the whole world, 'no original relation with nature unaltered, no living thing unaffected'.[38] He does not, however, discuss whether such a process is at all subject to limits imposed by the evolutionary history of the planet.[39]

Science as spectacle

The position of science in the set of types of knowledge is constantly under scrutiny. To some it is clearly the best kind of knowledge we have: it is convergent between different individuals and societies and can lead to powerful pieces of technology which distance us from those parts of nature we don't care for (like some bacteria) and bring us to the bits (like South Pacific islands) that we rather like. Those who claim that science is on the whole a bad thing are usually complaining about technology rather than science, which still has a glamorous image compared with engineering, for example.

The underlying question still seems to be, 'is there anything out there and if so, what is it like?'. All our sources of evidence seem to tell us that there is indeed something real beyond our skins and senses; that there is a world which would exist if we were not here to observe it. But what it is 'truly' like is another question since validating our models is not as easy as falling into circular arguments about facts validating theory when actually we have gathered those facts in accordance with a particular theory. No one believes that the mind is a clean slate upon which the senses inscribe their record of the world around us.[40] As both Kant and Nietzsche pointed out, experience is itself a species of knowledge which involves understanding and everything that reaches human consciousness is utterly and completely adjusted, simplified, schematised and interpreted. There is a special role, therefore, for the development of theories that are predictive since the forecasting of events that we ourselves are powerless to bring about provides an objective validation of the theory. That we cannot bring the event about ourselves is important, since otherwise there is the possibility of self-fulfilling prophecies.

Science has perhaps run out of models which easily fulfil these conditions. This is partly because it has dealt with most of the physical systems which seem to be largely determinate in their behaviour (like gravity for instance) and has got on the one hand into the sub-atomic level, where events are probabilistic in kind, and on the other into systems which contain life and which are therefore always open to indeterminate change. But the ability conferred by the computer to deal with large quantities of data has improved the predictability of indeterminate systems within the limits imposed by probability theory and the laws of chaos; elsewhere the kinds of experiment which would enhance the capability of science to deal with indeterminate systems like the biosphere usually involve doing something perceived as bad (like removing species from an ecosystem entirely or deforesting a watershed in an experimental catchment, or releasing genetically engineered organisms) and so are either not undertaken at all or else only in very restricted instances which may limit their predictive powers. Nevertheless, it is possible for philosophers of science to assert that experimentation leading to predictable change is the best test of scientific theory.[41]

This sort of discussion is conceptually rather pragmatic: in its chosen sphere, science is without doubt successful. In the spirit of this book, however, we ought to look a little into some of the arguments which constrain the notion of the primacy of science as a form of knowledge. One simple problem is the correspondence of the referent (such as an electron) to its description. In its purest form, science tells us of a referent which is described by a sender in terms of the proof of its truth. The sender then transmits the proof to an addressee (another scientist or a student) who accepts the statement. Such proof may be very difficult to find even outside the realm of human affairs. A complete definition of the initial state of a system which included, for example, all the independent variables would require an expenditure of energy at least as great as that consumed by the system to be defined. This being so, classical determinacy can work only within those systems which have at any rate conceivable (if not reachable) limits. Science also tells us that there are types of system which are by nature indeterminate, such as fractals and catastrophe surfaces, for instance. Here, even the stable form of the probability curve is not applicable. Lyotard quotes also the example of the real density of air in a sphere; as the volume of the sphere decreases, the density may be equal to 0 but may also be 10^n, where n is a large number.

Knowledge changes; what the physicists of the 1930s described as an electron is not what today's physicists would impute to that term. So we can construct a statement to the effect that just as no term used in the science of (say) 50 years ago referred, so it will turn out that no term used now, refers. P. B. Medawar goes on to quote the philosopher Tarski as saying that notions of truth and falsity are linguistic concepts for it is only sentences or propositions of which the truth or falsity can be affirmed.[42] Anything is then a statement about a statement delivered in a language in which we discourse about language which may then be termed a meta-language. The idea of 'the' scientific method is thus rendered particularly illusory, particularly since every recognition of what cannot be falsified (i.e. a provisional truth) is preceded by an imaginative preconception of what the truth might be, and this act is an act of creation or a *poesis*.[43] But the aim in the scientific testing of an idea is such that although the idea may begin as a story about a possible world, it ends by being, says Medawar, as nearly as we can make it, a story about real life.[44] It is then interesting to note that the great flagship of science is the controlled experiment and this is seen by some as a microcosm of the great project of the sciences, which is control of nature and of human societies as well.

It seems to follow that scientific terms are not the same as descriptions, which in turn leads to further difficulties in the relation between the referent and the observer. We can go even further with this type of analysis if we follow such writers as J.-F. Lyotard and P. B. Medawar when they distinguish the writing of science from its spoken form.[45] The latter, they state, is persuasive in nature rather than simply inductive, and tries to construct a narrative: in effect, to tell stories which have been scrupulously tested against real life.[46] Science's greatest achievement *in the real world* then becomes the having of new ideas. The world outside maps this view on to its own priorities and no longer asks 'is it true?' but 'what use are these new ideas?' and hence are they saleable? The convergence of many modern consumers of science (industry and government) upon this question is a noticeable feature of today's attitudes.

Science, though, is likely to get better at its tasks if it is allowed to work in an unfettered way, a condition which most societies do not tolerate except in time of war. Even in ideal conditions, though, there appears to be a consensus that science does not provide some kind of absolute knowledge of nature-in-itself; as Francis Bacon said, it is reserved for God and the Angels to be lookers-on. Science is an uncertain ally of environmental campaigners, for example, because it provides neither empirical nor episte-mological certainty.[47] This brings us immediately to the context in which it operates, in which its authority has been rational rather than charismatic or traditional. This means that now as never before the social context of science will mean that some will seek to harness it for political power rather than for disinterested empirical knowledge.[48]

Humanity in evolution

We are products of evolution at two scales: the cosmic and the organic. The first of these has set the limits of the parameters within which the second is nested: the laws of gravity and those of thermodynamics, for example. The second has produced the DNA molecule and then the species *Homo sapiens*. We can draw from this record both comfort (in the sense that we are at worst an accident rather than a pathology) and unease, that we do not fully realise the long-term implications of being products of such evolutionary histories.[49]

We can cast this discussion into sets of opposites which our evolution has produced. The first of these is the realisation that humans are both material and cultural beings. So much is obvious but may lead to the fallacy of assuming that one of these opposites is *de facto* dominant and perhaps even desirable. On the one hand this can lead us to the misconception of elevating the material to the point of primitivism: to where alternative groups want to go back to a pre-industrial and agrarian society which ignores any of the benefits of industrialisation. On the other hand, there is cultural dominance which assumes that whatever we can do, we ought to do and it is right for us to do it: a frontier or cowboy mentality in North American terms.

A second set of opposites can be discerned. That is, we are both emotional and rational creatures. Our nervous system seems to contain the capabilities for both rapid, instinctual response and calmer manipulation of symbols and abstractions. It is open to us to elevate one of these above the other, too. Dominance of the emotional can elevate wants to needs and luxuries to necessities and provide a 'rationale' for attachment to very high levels of production, no matter what the ecospheric consequences.[50] Concentration upon the rational (a common aim of, for example, universities) produces logical principles which are not accompanied by any commitment to action. (We all know we should take more exercise, eat less, be nicer to the cat, etc. but do we do it ... ?)

Third, we are both created and creators. We may not like the insignifi-cance shown to us by our role in energy flows (Table 6.1) or in views of the Earth from space but equally we may be unhappy with those places we can alter, especially if ecological instability is likely to happen. To be too much created is to adopt a fatalistic position in which we are the products of a simple set of chances and can in the long run do nothing: we are basically determined. The opposite is a kind of idealistic fallacy in which change is desirable in itself, no matter in what direction nor at what rate, even if the consequences are culturally destabilising for some, or indeed many.[51]

A major characteristic of western culture seems to be the adoption of one of each such pairs (there are others, of which 'good' and 'evil' is the most obvious) as an essential purpose of individuals and society so that ethics are directed at its triumph over its opposite. Dualistic-conquest thinking seems to have originated with Zoroastrianism, gained strength via Plato and other-worldly Christianity and found a congenial vehicle in the millennial thinking of the Puritans: American football, cowboy movies, Rambo, hi-tech mountain expeditions, prohibition and nineteenth century womens' corsets might all stand as outcomes. What we might examine more closely from the eastern traditions is the virtue of holding these pairs in a tension where both are seen to be necessary and where they never necessarily disappear. In the field of environmental construction, we might think of applying that to Cotgrove's two models (see Chapter 3) for instance.

Underlying any discussion of evolution in the last 4.2 million years is the thought that most of those years were spent as hunter-gatherers and that the behavioural patterns suited for survival in that way of life are not so adaptational to today's conditions. R. Ornstein and P. Ehrlich argue that our mental structure equipped us to live in an environment with many short-term challenges but with a stable backdrop.[52] Benefits had to be reaped quickly from day to day or month to month but the world renewed itself every year. The perceptual selectivity then exercised could ignore any longer term trends in the environment, they suggest, and that did not then matter given the short lives and low population density of the humans. Now it does, but we still do not perceive long-term threats and so are prone to the 'boiled frog' syndrome.[53] Even if we do not go very far with this set of ideas, we have to agree that cognitions of nature are inputs to world-views which do not deal directly with it, like free-market economics or the flows of capital.

Similar ideas are put into a more avowedly cultural setting in what A. Wilder calls 'context theory'.[54] Here, he contrasts 'the old ways' of A$\rightarrow\rightarrow$B linear causality, the notions of forces, atoms, closure, determinism, sameness, competition and short-range survival, with the 'new ways'. These are characterised by information, innovation, co-operation, communication, reciprocity, constraints, the nature of relationships, goalchanging and the capacity to use unexpected novelty in the service of long-range survival. So cultural evolution could move towards ideas such as that of society being part of nature and *vice versa* but noting that they are not the same parts, that communication is impossible without coding and the information in a system may be symbolic, imaginary or real. In the 'new ways' goal-seeking behaviour or teleonomy is different from a final end or *telos* and so is adaptatory rather than Utopian. What is not clear is how current debate about humanity–environment relations affects the change from one to the other ('Green' ideas are clearly part of the 'new ways') or whether indeed such transitions are taking place on anything but a very limited social and spatial scale.

THE TWO GREAT MODELS

From my study window at home, I can see the surface of the River Wear quite clearly. Opposite the house is a pool held up by an old weir, and its surface is sometimes broken by fish leaping out of the water momentarily, only to disappear again. In season there are two major species that do this: salmon and sea-trout. Likewise, I think that from the body of this book there appears from time to time one of two species of model of the world and our place in it which then drops back below the surface. By way of conclusion I reckon they must be hauled out for identification and indeed to see if they are edible or must be put back for someone else to catch another day. The two models of the world which in fact act as ground to much of the discussion in the various constructions relayed in this book are:

1 A model based on realism, one which acknowledges that there is a real world outside our bodies. This world can be affected by human actions but equally is likely to go its own way. Our part in it is small and perhaps accidental; certainly we are not a necessary component.
2 A model based on idealism, in which it is obvious that everything that we claim to know is a construct of our minds. Among other things this has come to give our species a dominance of natural processes and we might as well proceed on this basis: all is, eventually, possible.

The first leads us very noticeably through the values which have been deduced from a Darwinian view of organic evolution. These may be of the race-is-to-the-swiftest type; if our species is so dominant and manipulative then it is because it has passed some evolutionary test of fitness and has triumphed. We rule the Earth because we have learned the rules (such as, do not be too specialised) and constructed the best game plan. This is consistent with being seen as yet another species which is destined for extinction, as most have been, and perhaps in a cosmically short time rather than a long one. A metaphor for the first condition can be seen in one of Gary Larson's cartoons in which a row of insects are seated in a cinema: the title-frame of the movie is on the screen and it reads 'The Return of the Killer Windshield'.[55] A metaphor for the second is contained within the Gaia hypothesis discussed in Chapter 2. So evolutionary theory is not fundamentalist: its exegesis can lead in different directions.

The second model coincides with a basically humanistic view. Because of perceptual filtering (in which we receive possibly a trillionth of outside events[56]) everything we can claim to know is 100 per cent *Homo sapiens*. Since we can know in no other fashion then, we might argue, we have to accept our ecological dominance, achieved through the manipulation of symbols and of technology, and to use it 'wisely', though that is a self-referential concept without any absolute meaning. A metaphor for this might be the story of the Tibetan novice monk and his Abbot sitting on the hillside watching the prayer flags blowing in the wind. The novice asked the master, 'does the flag move or does the wind move?'. There was a long silence before the master replied, 'it is the mind which moves'. Any position of superiority does not of course preclude a voluntary relinquishment of it in favour of the construction of an ontological democracy where *Homo* is of equal value to, and has equal rights with, other animals, trees, rocks and the wind.

For the moment, I think (no doubt naively) that both are right. A modified realism appeals: that the cosmos exists in its own terms. Humans cannot know what it is like because of our perceptual inadequacies, even with science and technology to help us. The models we thus construct are imperfect and provisional. Hence, they may tell us different things (like those above) which have all the appearance of being correct or 'true' or even good. Explorers then find themselves in a world without a fixed set of signposts or other navigational aids: not for the first time. But I think that the fine arts are especially important here as they may presage movements and transformations in society which may shift environmental cognitions as well.

All coherence gone?

In his long poem of 1611, *An Anatomy of the World*, John Donne laments the breakdown of so much of the established order of things, cosmic and earthly, where 'new philosophy calls all in doubt', that he can say of the world:

> Tis all in pieces, all coherence gone
> All just supply and all relation: . . .
> For every man alone thinks that he hath got
> To be a phoenix and that then can be
> None of that kind, of which he is, but he.

For model two, the humanist one, this has a considerable resonance, for it is concerned with a fragmenting set of trends. In its context, today's humanists worry about the consequences of the abandonment of absolutes in human cognitions, values and structures. For many, this is at present focused on the lowered ambitions and decentred contingencies (i.e. the abandonment of a search for unity, totality and foundations) of the Post-modernists and their inheritance from Neitzsche of an untrammelled pluralism.

160

For them, model one (a world in which every other element has equal value in ethical and eventually perhaps in legal terms with humans and their constructions) cannot have coherence; to abandon the centrality of the human self is to court chaos and very likely the premature extinction of our species. Further, the ecological dominance of humanity now means that there is *de facto* no turning away from the responsibility, one way or another, of ordering the rest of the materials of planet Earth. And since humans live in two worlds at once, namely an ecological world and a psychological one, then the consequences of any failure to integrate these two worlds must be faced.

But in any case, we seem to be in the position crystallised by Balachandra Rajan (in a work of literary criticism[57] but clearly of relevance here):

> If the world and the experiencing self both change, both must be referenced to a point beyond change for a valid geometry of action to be constructed.

There seems no way out of that problem except by appeal to totalising fundamentalisms of either religious or secular varieties. These seem in general not to be helpful because they tend to be the products of previous historical patterns.[58] So there is a considerable problem lurking beneath this discussion, rather as the pool opposite my house might have a monster pike in its deeper parts. Suppose that both fact and value exist and are inter-subjectively knowable, but since they are so many, they often come into conflict with one another. Further, in many of the conflicts there is no standard by which to judge the competing claims, no measuring-rod against which to rank different approaches. Ultimately, there can be no convergence, in this view, on a universal civilisation with a unified and (we hope) harmonious relationship with the non-human, all undergirded by a rational morality. We are a highly inventive species and our common human traits underdetermine the forms of living (with all their implications for environmental change) that we can put together. Expectations of a convergence are therefore very likely misplaced. The implicit promise of the Enlightenment, namely that human society will eventually home in upon humanistic and liberal values, seems less likely than continued strife in which nature is likely to be one of the losing parties, as has happened frequently to date.[59] This has great consequences for constructions of human–environment relations which presuppose an end to which we might all look forward, even if it is a long way off. Instead, in this scenario, the pike eats the young ducklings and in turn is electro-fished by the Water Company.

Created and creating

The challenge is, therefore, to go forward as if more than one model were valid at any one time. We have perhaps no choice but to act like foxes even if we think like hedgehogs.[60] At a cosmic level, we are determined creatures: there is no escape from the major lineaments of matter and motion, thermodynamics and gravity, for instance. The Second Law prescribes the creation of entropy which we can only delay and put to good use *en route*, and also looses the arrow of time whose flight can go only one way. Chaos theory makes order in the cosmos more difficult to locate: all is a study of disequilibrium. To some extent this is redeemed by the idea that the creation of entropy is a source of intricacy: that irreversibility is a condition of the creation of complexity.

At the level of organic evolution we have choices: we can determine the course of future organic evolution by stamping our mastery upon this planet: by changing all the habitats, for example, or by interfering directly with the basic genetics of organisms. But we could also try co-evolution in which there is always space for other species, other habitats and where our own dominance is limited to spaces set aside for that: we can walk in the woods if we prefer them to the athletics stadium. This route means abandoning some pretty sacred cows to the slaughterhouse, though: ecological 'balance', a Golden Age, the idea of nature as totally produced by social forces.

But it is useless to pretend that all relationships must be between equals. The way of organic evolution is one of dominance at a local scale: the trees determine the ecology of the forest just as the density of the plankton determines the numbers of fish that eventually appear. The problem is in spatial and temporal scale: patterns of dominance which would pose no threat to any global process at a small scale can do so when they become worldwide. It then becomes a task for social theorists to find a stadium in which *Homo* can dominate and enjoy it without threatening the fabric of the streets and pubs around the ground.

Can we reasonably be asked to hold all these together? (By 'we' I think I mean the dominants of the world, who are at present the inhabitants of the western nations.) It is the action of holding that is perhaps the key here, for western thought is heavily infused with dichotomies or opposites such as *man and nature, reason and emotion, conquest and fatalism*, and finally, *good and evil*. So 'hard science' is better than 'emotionalism', 'being realistic' is superior to 'idealism' and always good must triumph over evil, preferably this week. We have either abandoned or never adopted those traits of thought from both West and East that have sought to balance both and to hold them in a creative tension where neither ever triumphs but neither ever goes away, either.[61] I suspect that the dichotomy between analysis and experience is critical here. Since the world of late classical antiquity, the West has separated these two, with the natural sciences becoming the main protagonists of *theoria*: being spectators and analysts. Yet their practitioners, like everyone else, ought to be moral agents. Thus to try to experience the world on its terms rather than as we have constructed it through rational presuppositions and socialised consciousness, may be a different route.[62]

At present, then, I argue that we have to know our place: to accept our place as it is presented to us from the past (all n millennia of it) and to make our abode as it is possible for us to do. Yet this latter can be in a spirit not of conquest but of holding. Heidegger wanted to transform 'I think therefore I am' into 'I care, therefore I am' and saw the great role of humankind as opening up a space for all else to have its Being. Given the state of the world, it looks as if contingency is all in this endeavour. That is, the world proceeds historically by the accumulation of small contingencies, with the great and incontrovertible cosmic laws of motion and matter in the background. The consequences, as S. J. Gould has argued,[63] include the realisation that individual events hold the power of transformation and so become a licence for us to participate in history. Thus most of us can continue to be foxes rather than hedgehogs, an easier pathway for many mortals, where getting the processes right could well be more liberating (for the whole planet) than setting up some long-term and overarching aim.[64] In the medium term this might, pragmatically, look like the calm formulations beyond the turbulence of the Great Climacteric suggested by Burton and Kates.[65] This begins with the abolition of nuclear arms and the reduction of other forms of armaments. It envisages a levelling-off of population growth rates everywhere and an evening-out of the distribution of wealth. Out of these might come a new environmental relationship, concerned with the long-term viability of the biosphere and a set of 'rights' for the natural environment. It seems inevitable, though, that environmental 'knowing' will have to be more unified: the kind of separation into science and non-science, different kinds of social sciences, dualism of reason and emotion, will be even more partial and unable to address the problems than at present. As a corrective, though, shifts into outright mysticism and largely unattainable senses of personal unity with the cosmos or nature may need a hard look.[66] But we would be moving towards Comenius's demand for knowledge that was 'universal, disgrac'd with no foul Casme'.

Humans, though, have the facility to see ahead and to worry about it and so it will never be enough for most of us to know our place and keep it. Rather like a concert, in the middle of a piece some of the audience is stealing covert glances at the programme to see what is coming next. Perhaps we can think of a new stage which goes beyond the yang of industrialism/empire and the yin of Green thought and politics to a creative tension beyond them. This will not be all rainbows and bulgur wheat yoghurt. But it might open space for yet further transformations of both humanity and nature which allow for the evolution of both, organically and culturally.

On the other hand, it may be an error to think of even a modest *telos* here since notions of a perfect whole are probably incoherent: there are great goods which are simply incompatible. The search for Utopia can become the tracking-down of an illusory final solution. Instead, then, priorities have never to be final and absolute, and equilibria are in constant need of repair. We will be doomed to choices and every choice may entail an irreparable loss; the dignity is in having the choice at all. This view sits rather well into the models of openness which may describe the world rather better than an equilibrium-based ecological world-view. We can only turn to a set of self-organising systems whose future we can influence in a contingent way, but the whole system is too complex to be foreseen: the search for Utopia must be abandoned in favour of a social theory for fudge.

So it is possible to see no virtue in a laid-down and prescriptive set of goals.[67] Let us rather go along with another pair of Buddhist monks, this time in the forests of Japan, although they could have been *mutatis mutandis* in Toledo or Tibet. A novice is walking along the forest path when by accident he meets the famous and awe-inspiring Abbot of another monastery. After bows, the novice, trembling, asks the Abbot, 'master, where is Buddha-nature to be truly found?'. The Abbot smiles and says, 'walk on; walk on'. It is the mind which moves.

NOTES

1 INTRODUCTION

1 Some would want the whole universe to be included as well; this presents a problem of degree not of kind.

2 Epistemology is the theory of knowledge: it seeks to define it, to differentiate its principal varieties, identify its sources and to place limits on it.

3 Some concepts of God would have Him/Her occupy this role but He/She seems not to produce many textbooks telling us exactly what the position is.

4 The pro-principle statement comes from H. Skolimowsky, *Living Philosophy. Eco-philosophy as a Tree of Life*, London: Arkana Books, 1992, p. 17; in its various forms it is subjected to considerable scrutiny by M. Midgley, *Science as Salvation. A modern myth and its meaning*, London and New York: Routledge, 1992.

5 From time to time, the term 'nature' will be used as a synonym for 'environment'. This too is a word with an immense range of meanings and usages: see some of the essays in A. O. Lovejoy, *Essays in the History of Ideas*, Baltimore: Johns Hopkins Press, 1948; R. Williams, *Keywords*, London: Flamingo Books, 1983, pp. 219–24. Unless the context militates otherwise, it will be used here for the non-human parts of the universe.

6 Ontology is the theory of what really exists as different from what appears to exist but in fact does not. But existence may apply to non-material phenomena, such as states of mind, in some philosophers' arguments.

7 J. Margolis, *The Truth about Relativism*, Oxford: Blackwell, 1991. The same book also uses the phrase 'adherence to the cognitive intransparency of the world'.

8 In some definitions, a construct is a useful fiction, i.e. it is acknowledged that there is nothing to which it corresponds in reality. At its extreme it leads to the situation described above where even the familiar terms of everyday language (tree, television, triple whisky) are merely constructs. But the dividing line between something which really exists and a useful fiction is distinctly imprecise. So we can develop the term **construction** to mean a coherent structure imposed by humans on the complexity of the world in order to make some sense of it. As such the construction will compose both objects accessible to the senses and the web of ideas that bind us and them together.

9 Remember that not all the living species on the Earth have been named. One estimate is that we can perceive about one trillionth of the forces which are present in the world about us.

10 Objective: 'belonging not to the consciousness or the perceiving or thinking subject, but to what is presented to this, external to the mind, real' (*Concise*

Oxford Dictionary, 7th corrected edn, 1983). This mode of enquiry which accepts that the observer can be completely detached from the observed is fundamental to natural science.

11 Subjective: 'belonging to . . . the consciousness or thinking or perceiving subject or ego as opp. real or external things' (*Concise Oxford Dictionary*, 7th corrected edn, 1983). We can refer to a similar epistemological vocabulary for the term **Positivism**, which involves first of all that objectivity is necessary and then aims to make law-like statements that describe and explain regularities in the relationship between phenomena. Contrast it with **Phenomenology** which starts out from human consciousness and tries to understand phenomena in that light without necessarily trying to formulate universal laws since human beings are unique individuals. But it integrates subject and object: man and nature form a continuous whole, for example; equally the world is held to have no objective existence outside the human mind. It is not a popular view with hard-line natural scientists.

12 Reductionism in philosophy is the process whereby statements are redefined in terms which are considered more elementary or more basic; in science it becomes the notion that a system can be understood in terms of the isolated operation of its component parts or sub-systems (*Concise Oxford Dictionary*, 7th corrected edn, 1983; *The Fontana Dictionary of Modern Thought*, eds A. Bullock and O. Stallybrass, London: Collins, 1977). For a detailed look at the implications of it in one field see A. R. Peacocke, 'Reductionism: a review of the epistemological issues and their relevance to biology and the problem of consciousness', *Zygon* 11, 1976, 307–34. He quotes Francis Crick, one of the discoverers of DNA, as saying, 'The ultimate aim . . . is in fact to explain *all* biology in terms of physics and chemistry.'

13 Holism is the thesis that wholes are more than the sum of their parts in the sense that they have characteristics that cannot be explained in terms of the properties and relations of the component parts. It considers that reductionism (*v.s.*) is a mutilation of whatever it is applied to (*The Fontana Dictionary of Modern Thought*, eds A. Bullock and O. Stallybrass, London: Collins, 1977).

14 The immediate derivation is from the Latin, *scire*, to know. On the Continent the split between the conventional subject matter of science and the humanities seems to be narrower: the words *sciences* (French) and *Wissenschaft* (German) can be used without strain in either branch of learning. In English-speaking nations, the natural sciences embrace fields like physics, chemistry, mathematics and geology with parts of psychology and geography.

15 There are, of course, scientists who believe that the methods of scientific discourse are the only ones which yield reliable information: all else is 'soft'. This view is called *scientism*. When Sir Keith (now Lord) Joseph was Minister of Education in the UK in the 1980s he forced the Social Science Research Council to rename itself the Economic and Social Research Council because he could not believe that its work was scientific.

16 For example A. P. A. Vink, *Landscape Ecology and Land Use* (ed. D. A. Davidson), London: Longman, 1983; Z. Naveh and A. S. Lieberman, *Landscape Ecology*, New York: Springer-Verlag, 1984; M. R. Moss (ed.), *Landscape Ecology and Management*, Montréal: Polyscience Publications, 1988; see the especially interesting contribution on methodology by Z. Naveh, 'Biocybernetic perspectives of landscape ecology and management', pp. 23–34. Although not entirely composed of contributions by geographers (but edited by one), an influential book in Geography has been W. L. Thomas (ed.), *Man's Role in Changing the Face of the Earth*, Chicago University Press, 1956. In this tradition see also

K. J. Gregory, *The Nature of Physical Geography*, London: Edward Arnold, 1985; *idem* (ed.), *Energetics of Physical Environment*, Chichester: Wiley, 1987. In Anthropology, see for example J. W. Bennett, *The Ecological Transition: Cultural Anthropology and Human Adaptation*, Oxford: Pergamon Books, 1976; E. F. Moran (ed.), *The Ecosystem Concept in Anthropology*, Boulder, CO: Westview Press, 1984, AAAS Selected Symposia 92.

17 The etymology is from Greek *oikos*, a household. 'Economics' also has this root.

18 J. E. Lovelock, *The Gaia Hypothesis. A New Look at Life on Earth*, Oxford: Oxford University Press, 1979. Lovelock is an independent chemist and an FRS: no product of the California commune culture. Gaia was the Greek name for the Earth Goddess. An updated and compact account of the theme is given by A. Watson, 'Gaia', in the *New Scientist's Inside Science* pullouts: **48**, 6 July 1991, 4 pp. J. R. Lovelock describes the genesis and evolution of the hypothesis in 'Hands up for the Gaia hypothesis', *Nature, Lond*, **344**, 1990, 100–02.

19 K. Lee, *Social Philosophy and Ecological Scarcity*, London: Routledge, 1989.

20 For a modern example of the combination of economics, politics and morals, see H. E. Daly, 'The economic growth debate: what some economists have learned but many have not', *J. Env. Econ. and Management* **14**, 1987, 323–36.

21 Deriving from the Greek *polis*, the city and hence the state, with the meaning of civic, i.e. of the city and state. The desire to formulate it as an objectively based study is shown by the frequent use of the term Political Science in academia.

22 The coming together is well illustrated by the book of H. Henderson, *The Politics of the Solar Age. Alternatives to Economics*, New York: Anchor Press/Doubleday, 1981.

23 See for example S. Cotgrove, *Catastrophe or Cornucopia: the Environment, Politics and the Future*, Chichester: Wiley, 1982; S. Cotgrove and A. Duff, 'Environmentalism, middle class radicalism and politics', *Sociology Review* **28**, 1980, 335–51.

24 In the first category see e.g. P. Hall, *Great Planning Disasters*, Harmondsworth: Penguin Books, 1981. In the second see H. H. Barrows, 'Geography as human ecology', *Ann. Assoc. Amer. Geogr.* **13**, 1923, 1–14, and R. Chorley, 'Geography as human ecology', in R. Chorley (ed.), *New Directions in Geography*, London: Methuen, 1973, pp. 155–70. Also the broader treatment by S. Macgill, 'Environmental questions and human geography', *Int. Social Sci. J.* **109**, 1986, 357–76.

25 P. L. Berger and T. Luckmann, *The Social Construction of Reality. A Treatise in the Sociology of Knowledge*, London: Allen Lane, 1967.

26 Edmund Husserl (1959–1938) was a German philosopher with whom is associated the crystallisation of this person-centred outlook, called **phenomenology**. For Husserl this was a necessary critique of the accepted methods of the natural sciences, which (he argued) ignored some basic truths about the individual and the reality of her/his perceptions of their total world.

27 A non-exhaustive list might include: writing, painting, photography, sculpture, music, weaving, fashion, flower arranging, drama and dance.

28 Friedrich Hebbel (1813–1863), quoted in W. H. Auden and L. Kronenberger (eds), *The Faber Book of Aphorisms*, London and Boston: Faber and Faber, 1964, p. 269.

29 J. Berger, *Ways of Seeing*, Harmondsworth: Pelican Books, 1972, is a case in point.

30 Concisely argued by R. Grove White, 'Mysteries in the global laboratory', *Times Higher Education Supplement*, 26 October 1990, 15.

31 Ethics is from the Greek *ethos*, nature or disposition; normative comes from the Latin *norma*, a carpenter's square.

32 How to find reference points for judgement is a perennial problem in ethics and philosophy. That there exists a transpersonal reality may be accepted. But we can have no idea what it is like since we lack the cognitive faculties to model it on a one-to-one correspondence. Thus we make models of it, using such data as we can acquire and such symbolic manipulations and inferences of and from the data as we can achieve.These constructions may well vary in their prescriptive as well as their 'factual' dimensions. But there is no absolute reference point against which they can be measured for their correspondence to reality. We have to accept that models A and B may tell us different things but seem to be of equal value: they are not obviously and consensually without value. A conflict of values, including the possibility of a conflict between 'good' things, is therefore to be expected. The thrust of the Enlightenment was that eventually people of liberal and humane views would come to a convergence. This may not after all be so.

33 See for example R. Attfield, *The Ethics of Environmental Concern*, Oxford: Basil Blackwell, 1983; T. Attig and D. Scherer (eds), *Ethics and the Environment*, Englewood Cliffs, NJ: Prentice Hall, 1983.

34 See J. D. Hughes, *American Indian Ecology*, El Paso, TX: Texas Western Press, 1983, and for one group in more detail R. K. Nelson, *Make Prayers to the Raven. A Koyukon View of the Northern Forest*, Chicago and London: University of Chicago Press, 1983.

35 One writer has talked of global cultural flow today as producing 'ethno-scapes, mediascapes, technoscapes, finanscapes, and ideoscapes': A. Appadurai, 'Disjuncture and difference in the global cultural economy', *Theory, Culture and Society* 7, 1990, 295–310.

36 Though who determines what any of us shall see is another and no less important matter.

37 See for starters, M. Sarup, *An Introductory Guide to Post-structuralism and Postmodernism*, New York: Harvester Wheatsheaf, 1988.

38 For this 'constructive' or 'revisionary' postmodernism, see D. R. Griffin (ed.), *The Reenchantment of Science: Postmodern Proposals*, Albany, NY: SUNY Press, 1988, especially pp. ix-xii.

39 W. Dilthey (1833–1911) writing in 1900, quoted in H. P. Rickman, *Wilhelm Dilthey. Pioneer of the Human Studies*, London: Elek Press, 1979, p. 88. Empirical accounts of human ability actually to change environments are demonstrated in a number of books such as I. G. Simmons, *Changing the Face of the Earth*, Oxford: Blackwell, 1989; B. L. Turner *et al.* (eds), *The Earth as Transformed by Human Action*, Cambridge: Cambridge University Press/Clark University, 1990; and at a more popular level (though by no means an unchallenging read), in C. Ponting, *A Green History of the World*, London: Sinclair-Stevenson, 1991.

40 C. J. Glacken, *Traces on the Rhodian Shore*, Berkeley and Los Angeles: University of California Press, 1967.

41 Nicely articulated by J. M. Roberts, *The Triumph of the West*, London: BBC Publications, 1985.

42 This the common translation of the German *Weltanschauung*. It signifies those elements of the lifeworld of an individual or society which are taken for granted. That of the West rests upon dualism of man and nature, linearity of time, the notion of progress, and the availability of technological solutions to problems as well as unlimited expectations of material prosperity.

43 M. Eliade, *The Myth of the Eternal Return or, Cosmos and History*, Princeton, NJ: Princeton University Press, 1971. Bollingen Foundation Series XLVI (first published in 1954).

44 A cosmogony was a recreation of the world. It might entail the sacrifice of a god-king or a ritual marriage between god-king and earth-goddess.
45 D. J. Herlihy, 'Attitudes toward the environment in medieval society', in L. J. Bilsky (ed.), *Historical Ecology. Essays on Environmental and Social Change*, London and Port Washington, NY: Kennikat Press, 1980, pp. 100–16.
46 Discussions by J. Passmore are helpful here: see his 'Attitudes to nature', in R. S. Peters (ed.), *Nature and Conduct*, London: Macmillan, 1975, pp. 251–64, and *Man's Responsibility for Nature*, London: Duckworth, 1980, 2nd edn.
47 A famous paper by the American historian Lynn White ('The historic roots of our ecologic crisis', *Science* **155**, 1967, 1203–07) ended by proposing St Francis as a patron saint for ecologists. Most of this paper has been attacked, and St Francis' ikon is still seen more often in Assisi than government offices and corporate boardrooms.
48 Not much limit to the literature, however. The high-profile beginning was D. Meadows *et al.*, *The Limits to Growth*, New York: Universe Books, 1972; a very full statement is P. Ehrlich, A. Ehrlich and J. P. Holdren, *Ecoscience*, San Francisco: Freeman, 1977, 2nd edn. A more concise account is in B. Ward, *Progress for a Small Planet*, Harmondsworth: Pelican Books, 1982, or in W. B. Clapham, *Human Ecosystems*, New York: Macmillan, 1981.
49 A large literature, but little of it global in scope. Apart from W. L. Thomas *op. cit.*, see my own *Changing the Face of the Earth*, Oxford: Blackwell, 1989; W. Holzner, M. J. A. Werner and I. Ikushima (eds), *Man's Impact on Vegetation*, The Hague: Junk, 1983; *Geobotany* 5; B. L. Turner *et al.* (eds), *The Earth as Transformed by Human Action*, Cambridge University Press and Clark University, 1991.
50 E. J. Dijksterhuis, *The Mechanization of the World Picture*, Oxford: Clarendon Press, 1961. See also A. W. Crosby, 'A Renaissance change in European cognition', *Environmental History Review* **14**, 1990, 19–32.
51 L. Winner, *Autonomous Technology*, Cambridge, MA: MIT Press, 1967; *idem*, *The Whale and the Reactor*, University of Chicago Press, 1986; J. Ellul, *The Technological Society*, New York: Knopf, 1964, all emphasise the 'autonomy' of technology once let out, genie-like, of the bottle.
52 R. H. Grove, 'The origins of environmentalism', *Nature, Lond*, **345**, 1990, 11–14.
53 J. B. Callicott and R. T. Ames (eds), *Nature in Asian Traditions of Thought*, Albany, NY: SUNY Press, 1989. But elaborations are beginning to appear, such as M. E. Tucker, 'The relevance of Chinese Neo-Confucianism for the reverence of nature', *Environmental History Review* **15**, 1991, 55–69.
54 A. Berque, *Le Sauvage et l'Artifice: Les Japonais devant la Nature*, Paris: Gallimard, 1986. For a more extended discussion see H. Tellenbach and B. Kimura, 'The Japanese concept of "nature"', in J. B. Callicott and T. T. Ames (eds) *Nature in Asian Traditions of Thought: Essays in Environmental Philosophy*, Albany, NY: SUNY Press, 1989, pp. 153–62; D. E. Shaner, 'The Japanese experience of nature', *op. cit.*, pp. 163–82.
55 Hwa Yol Jung, 'The ecologic crisis: a philosophic perspective, east and west', *Bucknell Review* **20**, 1972, 25–44; Shu-hsien Liu, 'Towards a new relation between humanity and nature', *Zygon* **24**, 1989, 457–68; M. E. Tucker, *op. cit.*, *supra*.
56 N. J. T. M. Needham *et al.*, *Science and Civilisation in China*, Cambridge University Press, 1954–6 vols by 1987, some with separate Parts.
57 O. Dwivedi, B. N. Tiwari and R. N. Tripathi, 'The Hindu concept of ecology and the environmental crisis', *Indian J. Public Administration* **30**, 1984, 33–67; G. J. Larson, '"Conceptual resources" in South Asia for "environmental ethics"', in Callicot and Ames, *op. cit.*, 1989, pp. 267–74.

58 Quoted in J. D. Hughes, *American Indian Ecology*, El Paso, TX: Texas Western Press, 1983. See also a special issue of *Environmental Review* (vol. 9, no. 2, 1985). For a contrast between native and European ideas, see C. Merchant, *Ecological Revolutions, Nature, Gender, and Science in New England*, London and Chapel Hill, NC: University of North Carolina Press, 1989. It has later emerged that some of these 'eco-friendly' statements have been more than a little enhanced in their translation.

59 Yi-Fu Tuan, 'Our treatment of the environment in ideal and actuality', *American Scientist* **58**, 1970, 244–9.

60 G. A. Klee (ed.), *World Systems of Traditional Resource Management*, London: Edward Arnold, 1980.

61 A successful (but not entirely so) London football team.

62 J. R. Short, *Imagined Country: Society, Culture and Environment*, London and New York: Routledge, 1991, p. xvi.

63 And especially the moral virtues that flow from these places. As in North America, the small family farm is seen as a well of virtue, even though the water may occasionally be a little cloudy. See the TV adaptations of James Herriot's stories about a Yorkshire veterinary practice.

64 N. Luhmann, *Ecological Communication*, Cambridge: Polity Press, 1989, trans. J. Bednarz (first published in German in 1986). This is very much a realist view, informed by natural science. There are other opinions which suggest that direct and intuitive cognition of 'nature' by humans is possible; this is especially true of the 'New Age' or 'Aquarian' groups which regard Gaia as something more than a scientifically based metaphor.

65 This may of course be general: for the globe as a whole, or for parts of it, or largely for our own familial descendants. Look at the dedications of books on environmental topics and see how often children and children's children get mentioned.

66 G. T. Geballe, 'The environment is dead', *Society and Natural Resources* **3**, 1989, 11–14.

67 Quoted in J. E. Saddler, *J. A. Cornelius and the Concept of Universal Education*, London: Allen and Unwin, 1966.

2 THE NATURAL SCIENCES AND TECHNOLOGY

1 'Science' derives from the Latin *scire*, 'know'; 'technology' from the Greek *techne*, 'art, craft, skill', possibly traceable to the Indo-European root, *teks*, 'weave, fabricate', cf. 'textile'.

2 For the 'calmer' view see for example J. Ziman, *Reliable Knowledge: An Exploration of the Grounds for Belief in Science*, Cambridge: Cambridge University Press, 1978; *idem, Public Knowledge. An Essay Concerning the Social Dimension of Science*, Cambridge: Cambridge University Press, 1968. Both seem well-balanced, are clear and not without humour. Another useful beginning would be A. F. Chalmers, *What is This Thing Called Science?*, Milton Keynes: Open University Press, 1982. 'Anti' books are referenced at appropriate points.

3 For a summary, see J. Needham, *The Grand Titration. Science and Society in East and West*, London: Allen and Unwin, 1969. On the fundamentals of Chinese science, see vol. 2 (*History of Scientific Thought* (1956)) of the magisterial series by J. Needham and collaborators, *Science and Civilisation in China* (Cambridge University Press 1954–).

4 The classical method of positivist science proceeds by the statement of hypotheses, laws and theories:

● **Hypotheses** are statements not yet accepted as laws or as true. Positivist methodology aims to verify or falsify these statements.

- **Laws** are statements that are true in all times and places and therefore can lead to temporal and spatial prediction.
- **Theories** are unified and interrelated systems of laws with explanatory force. They often structure knowledge.

What is sometimes called 'the scientific method' consists therefore of observation and classification of data received through the senses (or instrumented extensions of them), followed by the formation of a hypothesis. If this is verified by experiment or by the further accumulation of data then it receives the status of a law. A body of these laws which contains explanatory inferences constitutes a theory. This methodology is especially suitable for experimental sciences and for closed systems but is less helpful in historical sciences like ecology and palaeontology, and in the social sciences. Historical sciences use comparative detail and observational richness as foundations. It is not possible to see a past event directly but inference is common to all the sciences: nobody has seen a black hole. Charles Darwin, we may note, rejected the widely held notion that a cause must be seen directly in order to qualify as a scientific explanation. He accepted that independent sources might come together to provide a particular historical pattern and used the word 'consilience' ('jumping together') for such a process.

5 But our knowledge of the external world must be bounded by what the human organism is capable of receiving by way of messages from it. For this reason, Ziman (*op. cit.*, 1978) prefers the term 'intersubjective' to 'objective'.

6 A. F. Chalmers, *Science and its Fabrication*, Milton Keynes: Open University Press, 1990.

7 Technology is usually science-based but does not *have* to be: it could in theory be based on beauty, or on the preferences of the manager of the England football team.

8 This important idea is discussed (in its general form) in Chalmers, *op. cit.*, 1982, and in most books on the nature of science. See, for instance, M. Hesse, *Revolutions and Reconstructions in the Philosophy of Science*, Brighton: Harvester Press, 1980. This is an especially helpful book because it deals with the social sciences as well.

9 Concorde is discussed in P. Hall, *Great Planning Disasters*, Harmondsworth: Pelican Books, 1981; the lack of introspection about the dropping of the Bomb at Hiroshima and Nagasaki in L. Giovanetti and F. Freed, *The Decision to Drop the Bomb*, London: Methuen, 1967.

10 L. Winner, *Autonomous Technology*, Cambridge, MA: MIT Press, 1977; *idem*, *The Whale and the Reactor*, Chicago and London: University of Chicago Press, 1986.

11 The classic polemic of this case is J. Ellul, *The Technological Society*, New York: Vintage Books, 1964 (first published in Paris, 1954).

12 J. Cherfas, 'Ecology invades a new environment', *New Scientist* **116** (1583, 22 October 1987), 42–6.

13 On the relations of theory in ecology and evolution, see G. E. Hutchinson, *The Ecological Theatre and the Evolutionary Play*, New Haven, CT: Yale University Press, 1965.

14 *Ibid.*

15 In R. P. McIntosh, *The Background of Ecology: Concept and Theory*, Cambridge: Cambridge University Press, 1985, p. 321.

16 R. P. McIntosh, *op. cit.* See also D. Worster, *Nature's Economy: A History of Ecological Ideas*, Cambridge: Cambridge University Press, 1985. McIntosh criticises this book on the grounds that it deals with environmental thought rather than ecological ideas *sensu stricto*.

17 E. Mayr, *The Growth of Biological Thought: Diversity, Inheritance and Evolution*, Cambridge, MA and London: The Belknap Press of Harvard University Press, 1982.

18 The term *autogenic* is usually applied, contrasting with *allogenic*, meaning influenced from 'outside' e.g. by climate or humans.

19 The key paper proposing a new set of models for succession is W. H. Drury and I. C. T. Nisbet, 'Succession', *J. Arnold Arboretum* 54, 1973, 331–68, and a book which enlarges on the ideas, S. T. A. Pickett and P. S. White (eds) *The Ecology of Natural Disturbance and Patch Dynamics*, Orlando, FL: Academic Press, 1985. It is fair to say that proponents of equilibrium ideas are not silent, though they are less in favour of determinism than before.

20 F. N. Egerton, 'History of ecology: achievements and opportunities. Part I', *J. Hist. Biol.* 16, 1983, 259–310.

21 For the transition of ecological science from equilibrium-orientation to open-endedness, see D. Worster, 'The ecology of order and chaos', *Environmental History Review* 14, 1990, 1–18.

22 W. A. Reiners, 'Disturbance and basic properties of ecosystem energetics', in H. A. Mooney and M. Godron (eds) *Disturbance and Ecosystems: Components and Response*, New York: Springer-Verlag, Ecological Studies 44, 1983, pp. 83–98.

23 E. P. Odum, *Basic Ecology*, Philadelphia: Saunders, 1983; H. T. and E. C. Odum, *Energy Basis for Man and Nature*, New York: McGraw Hill, 1975.

24 See the numerous books, chapters and papers by biologists, geographers and others with titles like 'Man and ecosystems' or 'Human impact upon ... ecosystems'.

25 See e.g. R. J. Chorley, 'Geography as human ecology', in R. J. Chorley (ed.) *New Directions in Geography*, London: Methuen, 1973, pp. 155–70.

26 W. Bennett, *The Ecological Transition. Cultural Anthropology and Human Adaptation*, Oxford: Pergamon Books, 1976.

27 E. O. Wilson, *Sociobiology: the New Synthesis*, Cambridge, MA: Harvard University Press, 1975; P. Colinvaux, *The Fates of Nations: A Biological Theory of History*, New York: Simon and Schuster, 1980/Harmondsworth: Penguin Books, 1983.

28 G. W. Barrett and R. Rosenberg (eds), *Stress Effects on Natural Ecosystems*, Chichester: Wiley, 1981.

29 S. I. Auerbach, 'Ecosystem response to stress: a review of concepts and approaches', in Barrett and Rosenberg, *op. cit.*, 1981, pp. 29–41.

30 G. P. Marsh, *Man and Nature, or Physical Geography as Modified by Human Action*, New York: Scribner, 1864; new edition (ed. D. Lowenthal), Cambridge, MA: Belknap Press of Harvard University Press, 1965.

31 G. H. Orians *et al.*, 'Dealing with uncertainty', in G. H. Orians (ed.) *Ecological Knowledge and Problem-Solving: Concepts and Case Studies*, Washington, DC: National Academy Press, 1986, pp. 88–92.

32 Quoted in R. P. McIntosh, *op. cit.*, 1985, p. 311.

33 P. Wathern (ed.), *Environmental Impact Assessment: Theory and Practice*, London: Unwin Hyman, 1988.

34 These scenarios were hypotheses based on models. These models are under continuous refinement and some show a wider range of outcomes than the original publications suggested, not surprisingly. It is scarcely useful to cite the latest work available at the time of submission of this manuscript: look in the *Bulletin of the Atomic Scientists* for the latest developments. The events of 1989 in the USSR and Eastern Europe may have made this type of work seem less imperative; I'm not so sure.

35 A. Henderson-Sellers, 'Climate, models and geography', in B. Macmillan (ed.) *Remodelling Geography*, Oxford: Blackwell, 1989, pp. 117–46.

36 An economic model of world energy use with a few emissions coefficients is no longer enough. GCMs must now encompass the emissions of methane from livestock, land use changes and rice cultivation, the output of nitrogen oxides from various sources, the success of industrial chemistry in coping with the Montreal convention on CFCs, and the spatial *and* temporal cycles of emission of all these Radiatively Important Gases (RIGs). See M. Scott *et al.*, 'Global energy and the greenhouse effect', *Energy and Environment* 1, 1990, 74–91.

37 A letter from Wallace to Darwin in 1858 probably precipitated Darwin into the publication of *On the Origin of Species* in 1859.

38 Some argue that we are the finest product of this process but it is a rather circular argument and in any case the game is not yet over: it is too early to nominate the player of the match.

39 This definition comes basically from *The Fontana Dictionary of Modern Thought* (eds A. Bullock and O. Stallybrass), London: Fontana/Collins, 1977.

40 D. Worster, *Nature's Economy: A History of Ecological Ideas*, Cambridge: Cambridge University Press, 1985.

41 There are numerous reprints of all these books, in anything from burnished unicorn-hide to paper virtually free of any type of fibre. The details of the original editions are: *On the Origin of Species by Means of Natural Selection, or the Preservation of Favoured Races in the Struggle for Life*, London: John Murray, 1859 [price 15 shillings]; *The Descent of Man and Selection in Relation to Sex*, London: John Murray, 1871, 2 vols [price £1–4–0]. It was in this work (vol. I, p. 2) that the word 'evolution' occurs for the first time in Darwin's works. *The Beagle* material appeared from 1838–43, published in London by Smith, Elder and Co. The *Zoology*, for example, was in 19 numbers (5 parts), a total of 617 pages and priced at £8–15–0. See R. B. Freeman, *The Works of Charles Darwin: An Annotated Bibliographical Handbook*, London: Dawsons, 1965.

42 S. J. Gould, *Wonderful Life: The Burgess Shale and the Nature of History*, London: Hutchinson, 1989, p. 35.

43 There is a full discussion of this in Worster, *op. cit.* I have not gone to all the Victorian authors to find out exactly what they mean by civilisation. A combination of clean water, good sewers, *Bradshaw*, cheap servants, cricket and Christianity seems to be the sort of ideal.

44 Worster, *op. cit.*, p. 177. The phrase has a curious echo of the domino theory to it and was as about as successful as US policy in Southeast Asia in the 1960s and 1970s. Worster also quotes Huxley as being very pleased that the native flora of Tasmania, for example, was being ousted by *English* plants.

45 For a full discussion of this and other matters related to evolution see the eminently readable Mary Midgely, *Evolution as a Religion: Strange Hopes and Stranger Fears*, London: Methuen, 1985.

46 Another quote from Francis Crick, reproduced by Midgley, *op. cit.*, p. 57. In 1929, J. D. Bernal apparently thought that scientists *per se* would become a new species. Note also that some biologists dissent from the Darwinian view that the species is the focus of selection by environmental pressures: for them **group selection** is more powerful. See the work of V. C. Wynne-Edwards, as in *Evolution through Group Selection*, Oxford: Blackwell Scientific, 1986; *idem*, 'Ecology denies neo-Darwinism', *The Ecologist* 21, 1991, 136–41.

47 T. Ingold, *Evolution and Social Life*, Cambridge: Cambridge University Press, 1986, discusses all these views.

48 Of which the clearest and most succinct is M. Midgley, *op. cit.*

49 E. Jantsch, *The Self-Organizing Universe: Scientific and Human Implications of the Emerging Paradigm of Evolution*, Oxford: Pergamon Press, 1980.

50 N. Georgescu-Roegen, 'Energetic dogma, energetic economics and viable technologies', *Adv. Econ. Energy and Resources* 4, 1982, 1–39.

51 I. Prigogine and I. Stengers, *Order out of Chaos: Man's New Dialogue with Nature*, London: Flamingo Books, 1985.

52 N. Georgescu-Roegen, 'Feasible recipes versus viable technologies', *Atlantic Economic Journal* 12, 1984, 21–31.

53 J. E. Lovelock, *Gaia: A New Look at Life on Earth*, Oxford: Oxford University Press, 1979; updated in *idem*, 'Gaia: the world as living organism', *New Scientist* 112, 1986, 25–8 and in popular form in 'Stand up for Gaia', *Biologist* 36, 1989, 241–7. The latest full account is J. E. Lovelock, *The Ages of Gaia: A Biography of our Living Earth*, Oxford: Oxford University Press, 1988. A concise and non-technical discussion of the ideas (and some of the difficulties associated with their acceptance) can be found in *The Economist*, 22 December 1990, 113–19.

54 D. Lindley, 'Is the earth alive?', *Nature, Lond*, 332, 1988, 483–4.

55 'Nothing yet suggests the jump is imminent . . . we should be like those doctors of old who could not cure people but tried at least not to harm them.' J. E. Lovelock, quoted in *The Guardian*, 30 December 1988. See also J. E. Lovelock, 'Planetary medicine. Stewards or partners on Earth?', *Times Literary Supplement* (4615, 13 September 1991), 7.

56 cf. N. Georgescu-Roegen, *op. cit.*, 1982, 34, 'Perhaps the destiny of man is to have a short, but fiery, exciting and extravagant life rather than a long, uneventful and vegetative existence. Let other species – the amoebas, for example – which have no spiritual ambitions inherit an earth still bathed in plenty of sunshine.'

57 There are many anti-Gaia views. A concise statement is M. R. Hoare, 'When the earth moved', *Times Higher Education Supplement*, 1 February 1991, 15.

58 The journal *The Ecologist* has constantly criticised the Darwinian construction of the changes in life on earth and finds support in the Gaia notion. See P. Bunyard and E. Goldsmith (eds) *Gaia and Evolution*, Camelford, Cornwall, UK: Wadebridge Ecological Centre, 1989. They also suggest that Gaia herself and not the species is the unit of evolution.

59 Various interesting data, some with an environmental occupation slant, can be seen in D. E. Morgan, 'Living with risk', *Biologist* 36, 1989, 117–24.

60 These data and the categories of perceived risk are from F. Warner (ed.) *Risk Assessment*, London: The Royal Society, 1983. See also J. Conrad, *Society, Technology and Risk Assessment*, London: Academic Press, 1980; W. D. Rowe, *An Anatomy of Risk*, New York: Wiley, 1970.

61 An influential book on risk makes the point that advances in scientific knowledge have in many instances expanded the universe of things about which we cannot speak with confidence. J. Douglas and A. Wildavsky, *Risk and Culture. An Essay on the Selection of Technological and Environmental Dangers*, Berkeley and Los Angeles: University of California Press, 1982.

62 See, for example, R. E. Kasperson *et al.* (eds), *Understanding Global Environmental Change. The Contributions of Risk Analysis and Management*, Clark University: ET Program, 1990.

63 I. Stewart, *Does God Play Dice? The Mathematics of Chaos*, Oxford: Blackwell, 1989. By **stochastic** is meant a process involving a random variable, the successive values of which are not independent. The Greek origin of the word involves the notion of guessing, though eventually *stokhos* ('target') is involved and so the freedom to be variable is likewise circumscribed.

64 In his book *Chaos: the Making of a New Science* (New York: Viking, 1987),

J. Gleick discusses the work in population ecology of R. M. May, e.g., 'Simple mathematical models with very complicated dynamics', *Nature, Lond,* **261**, 1976, 459–67; see also W. H. Schaeffer, 'Order and chaos in ecological systems', *Ecology* **66**, 1985, 93–106. A broader interpretation of chaos and its relevance for the general open-to-the-future condition of the Universe is offered by P. Davies, 'Chaos frees the universe', *New Scientist* **128** (1737, 6 October 1990), 48–52.

65 J. M. Diamond, 'Human use of world resources', *Nature, Lond,* **328**, 1987, 479–80.

66 A. Leopold, *A Sand County Almanac,* Oxford University Press, 1949, is probably his most famous expression of these ideas. There is a biography of him: S. L. Flader, *Thinking Like a Mountain: Also Leopold and the Evolution of an Ecological Attitude towards Deer, Wolves and Forests,* Columbia, MO: University of Missouri Press, 1974. A discussion of Leopold's ideas in a broader context of environmental philosophy occurs in E. C. Hargrove, *Foundations of Environmental Ethics,* Englewood Cliffs, NJ: Prentice Hall, 1989, especially chapter 5.

67 P. Colinvaux, *The Fates of Nations. A Biological Theory of History,* New York: Simon and Schuster, 1980/Harmondsworth: Pelican Books, 1983, p. 253. Actually, the quotation is from an Appendix which is a reprint of the Fourth Tansley Lecture given in 1981.

68 J. R. Vallentyne, 'The necessity of a behavioural code of practice for living in the biosphere, with special reference to an ecosystem ethic', in N. Polunin (ed.) *Ecosystem Theory and Practice,* Chichester: Wiley, 1986, 406–14.

69 Summarised by D. Worster, 'Ecology of order and chaos', *Environmental History Review* **14**, 1990, 1–18.

70 A. Brennan, *Thinking about Nature: An Investigation of Nature, Value and Ecology,* Athens, GA: University of Georgia Press, 1988.

71 K. Lee, *Social Philosophy and Ecological Scarcity,* New York and London: Routledge, 1989.

72 N. Luhmann, *Ecological Communication,* Cambridge: Polity Press, 1989 (first publ. in Germany 1986; trans. J. Bednarz).

73 There are extended versions of these last few ideas in the contributions of H. Morowitz ('Biology of a cosmological science', 37–49) and J. B. Callicott ('The metaphysical relations of ecology', 51–64) to J. B. Callicott and T.T. Ames (eds), *Nature in Asian Traditions of Thought,* Albany, NY: SUNY Press, 1989.

74 Science seems to learn to read the text of nature and to order the stimuli received in an acceptable way but there are perhaps no techniques for reading all the parts of the text equally well. Those for predictability and control, though, are highly prized in both intellectual and practical terms. Some scientists would argue that predictability is a better criterion of good science than e.g. intersubjective (= 'objective') knowledge: if Robinson Crusoe had developed a good science of botany (before the arrival of Man Friday) then it would have enabled him to predict where edible plants were to be found.

75 See my *Changing the Face of the Earth,* Oxford: Basil Blackwell, 1989. The reference is to Jarrow monastery on Tyneside in the first millenium AD where it is said that coal as well as wood was used for heating.

76 P. Scott, 'Knowledge's outer shape, inner life', *Times Higher Education Supplement,* 16 August 1991, 12.

77 P. B. Medawar, *The Art of the Soluble: Creativity and Originality in Science,* Harmondsworth: Penguin, 1969. Medawar was a distinguished biologist and champion of science and was certainly not out to detract from its achievements.

78 H. Longino, *Science as Social Knowledge,* Princeton, NJ: Princeton University Press, 1990.

79 The 'web of belief' idea is found in W. V. Quine and J. S. Ulliam, *The Web of Belief*, New York: Random House, revised edn, 1978. It is put into a wider context in a book claimed to be general in approach but in fact pretty technical in places: J. J. C. Smart, *Our Place in the Universe: A Metaphysical Discussion*, Oxford: Basil Blackwell, 1989.

80 D. Faust, *The Limits of Scientific Reasoning*, Minneapolis: University of Minnesota Press, 1984.

81 C. Mitcham and R. Mackey (eds), *Philosophy and Technology*, Glencoe, IL: Free Press, 1983. The term 'hydra-headed' itself may not be actually used in that book but is common in polemical literature: it refers to the monster of Greek legend which had nine heads. If one was struck off, then it was replaced by two new ones.

82 This quotation and the paragraph in which it is embedded come from M. Adas, *Machines as the Measure of Men: Science, Technology and Ideologies of Western Dominance*, Ithaca, NY and London: Cornell University Press, 1989. *A propos* the Revd J. Cummings, we might remark that Hinduism is still alive and flourishing in India, and elsewhere.

3 SOCIETY AND ENVIRONMENT: MATERIALIST INTERPRETATIONS

1 In philosophy, realism assumes that there is a world of material things whose existence does not depend on an observer's mind being aware of them. (In common usage of course it usually reflects the routing of political power: 'let's be realistic about this' means something bad for the persons or natural objects on the receiving end.) Materialism is more explicit in assuming that everything that exists occupies some volume of space for some period of time and is accessible to senses like sight and touch. It denies existence to e.g. minds and mental states unless they are identified with the biochemistry of the brain and nervous system. It excludes therefore disembodied minds like those of the dead or the unborn, or the notion of God.

2 This usage is derived from T. S. Kuhn, *The Structure of Scientific Revolutions*, Chicago: University of Chicago Press, 1962, in which he proposed that there was at any one time a dominant thought-pattern in any science which ruled even the questions that could be asked: 'the entities that nature does and does not contain' (p. 108). In time, evidence accumulated which overthrew that pattern and established a new one. An example would be the way in which the determinism of Newtonian physics was succeeded in the 1930s by the more probabilistic quantum mechanics pioneered by Einstein and others. Outside physics, the idea has possibly never applied quite so well but that has not prevented its widespread use. Thus we come to employ it loosely, as here, to mean a structured and coherent thought- and practice-pattern identifiable within a discipline ('In learning a paradigm, the scientist acquired theory, methods and standards together, usually in an inextricable mixture', Kuhn, *op. cit.*).

3 S. Sarkar, *Green-Alternative Politics in West Germany, Vol. II, The Greens*, Tokyo: United Nations University Press, 1991.

4 But it is very difficult for ecologists and economists to find a common language. See the non-meeting of minds in for example D. O. Hall, N. Myers and N. S. Margaris, *Economics of Ecosystem Management*, Dordrecht, Boston and Lancaster: Junk, Tasks for Vegetation Science 14, 1985.

5 There are numerous books on the economics of environment and resources, with the two not always differentiated, as in T. Teitenberg, *Environmental and*

Natural Resource Economics, Glenview, IL: Scott, Foresman, 1984. An introduction for non-specialists is A. Cottrell, *Environmental Economics*, London: Edward Arnold, 1978; a more thorough and theory-based treatment is A. V. Kneese, *Economics and the Environment*, Harmondsworth: Penguin, 1977. On resources, there is G. A. Norton, *Resource Economics*, London: Edward Arnold, 1984, and the much more mathematical P. Dasgupta, *The Control of Resources*, Oxford: Basil Blackwell, 1982. There are many others, especially those emanating from the Resources for the Future think-tank in Washington, DC.

6 See for example R. Lecomber, *The Economics of Natural Resources*, London: Macmillan, 1979.

7 B. M. Hanlon, 'An energy standard of value', *Ann. Amer. Acad. Soc. and Pol. Sci.* **405**, 1973, 139–53.

8 Discounting is the act of reducing a future value to a present equivalent by dividing the future value by the discount factor $(1 + r)^t$ where r is the annual discount rate and t is the number of years into the future. An indefinite stream of future values (good and bad) is reduced to a finite present value. It is a contentious question as to how the rate should be calculated and indeed whether the environment should be discounted in the same way as the material products of the economy. See e.g. C. Price, 'Project appraisal and planning for overdeveloped countries', *Env. Management* **8**, 1984, 221–42, and the extended discussion in D. W. Pearce, A. Markandya and E. B. Barbier, *Blueprint for a Green Economy*, London: Earthscan Publications, 1989, ch. 6.

9 See D. W. Pearce, *Cost-Benefit Analysis*, London: Macmillan, 1983, 2nd edn; *idem*, *Economic Values and the Natural Environment*, London: UCL Discussion Papers in Economics 87–8, 1978.

10 D. Pearce *et al.*, *op. cit.*, 1989.

11 And indeed, accounting generally: one collection of essays discusses the problems of company reports and their treatment of environmental values. See D. Owen, *Green Reporting: Accountancy and the Challenge of the Nineties*, London: Chapman and Hall, 1991.

12 One author argues that CBA *cannot* identify the best paths for change: that it is basically a device for supplying politicians with numbers. See G. Maier-Rigaud, *Ecological Economics and Global Change*, Bonn: Institut for Europäische Umweltpolitik, 1991.

13 D. W. Pearce, 'Optimal prices for sustainable development', in D. Collard *et al.* (eds) *Economics, Growth and Sustainable Environments, Essays in Memory of Richard Lecomber*, London: Macmillan, 1988, 57–83.

14 D. W. Pearce *et al.*, *op. cit.*, 1989, ch. 2.

15 For a discussion of this, see W. U. Chandler, *The Changing Role of the Market in National Economies*, Washington, DC: Worldwatch Paper 72. For a rejection of much of classical free-market economics in terms of the actual behaviour of consumers in making choices, see P. Earl, *The Economic Imagination: Towards a Behavioural Analysis of Choice*, Brighton: Wheatsheaf Press, 1983.

16 P. Worsley, *Marx and Marxism*, Chichester: Ellis Horwood/London and New York: Tavistock Publications, 1982.

17 A. Schmidt, *The Concept of Nature in Marx*, London: New Left Review, 1971, trans. B. Fowkes (German original publ. 1962).

18 H. L. Parsons (ed. and comp.), *Marx and Engels on Ecology*, Westport, CT and London: Greenwood Press, Contributions in Philosophy no. 8, 1977. The volume of Marx's work in particular makes it a bit like the Bible: search hard enough and a phrase can be found to support a particular position.

19 M. Quaini, *Geography and Marxism*, Oxford: Basil Blackwell, 1982, ed. R. King (first publ. in Italian in 1974).

20 This is in the *Economic and Philosophical Manuscripts* of 1844, which contain many of the more environmentally sensitive remarks by Marx.

21 Four papers in the journal *Environmental Ethics* encapsulate some of the arguments: a critical view is D. C. Lee, 'On the Marxian view of the relationship between man and nature' (**2**, 1980, 3–16); answered by C. Tolman, 'Karl Marx, alienation and the mastery of nature' (**3**, 1981, 63–74), and by V. Routley, 'On Karl Marx as an environmental hero' (**3**, 1981, 237–44). Hwa Yol Jung tries to go beyond both in 'Marxism, ecology, and technology' (**5**, 1983, 169–71).

22 The critical theory of the Frankfurt School has taken these ideas further: see e.g. H. Marcuse, *One-dimensional Man*, Boston: Beacon Hill Press, 1964; *idem*, *Counterrevolution and Revolt*, Boston: Beacon Hill Press, 1969. The latter is especially strong on the humanity–nature relationship.

23 'Everything else being equal, it can be said that the ecological problem in the USSR and other socialist countries is not as bad as in the capitalist world simply because production and consumption are not as great': J. O'Connor, 'Political economy of ecology of socialism and capitalism', *Capitalism Nature Socialism* **3**, 1989, 93–107. This does rather beg the question of what else is equal and certainly seems not to be the majority view after the opening up of these nations in the late 1980s.

24 J. R. L. Proops, 'Ecological economics: rationale and problem areas', *Ecological Economics* **1**, 1989, 59–76.

25 This is the theme of much of A-M. Jansson (ed.), *Integration of Economy and Ecology. An Outlook for the Eighties*, Stockholm: University of Stockholm Askö Laboratory, 1984. The notion of energy as a mediator between ecology and economics has a history dating back at least to the 1880s. See J. Martinez-Alier, *Ecological Economics: Energy, Environment and Society*, Oxford: Blackwell, 1987.

26 See for example R. B. Noorgard, 'Coevolutionary development potential', *Land Economics* **60**, 1984, 160–73. This is mostly an account of co-evolution in biology, with the metaphor transferred to the economics of non-renewable resources. There are some very interesting remarks on the development of models in economics as responses to the real world's patterns of trade and technology; he suggests that a new co-evolutionary type of model should now be developed. In another work (R. R. Nelson and S. G. Winter, *An Evolutionary Theory of Economic Change*, Cambridge, MA and London: Belknap Press, 1982) 'evolutionary' has the connotation of 'non-equilibrium' rather than any biological overtones. This theme of the irreversibility of time and the lack of equilibrium constantly introduced by technological change is also dealt with in C. Perrings, *Economy and Environment: A Theoretical Essay on the Interdependence of Economic and Environmental Systems*, Cambridge: Cambridge University Press, 1987.

27 M. J. Lavine, 'Fossil fuel and sunlight: relationship of major sources for economic and ecological systems', in A-M. Jansson (ed.) *Integration of Economy and Ecology – an Outlook for the Eighties*, Stockholm: University of Stockholm Askö Laboratory, 1984, 121–51.

28 The image is that of a water-lily whose leaf-area doubles every day. On the twenty-ninth day it has merely covered half the pond but of course on the next day it has overlaid the whole pond. The metaphor can relate to kinds of growth rates, e.g. of human populations, energy use, metal ore consumption and so on. See L. R. Brown, *The Twenty-Ninth Day: Accommodating Human Needs and Human Numbers to the Earth's Resources*, New York: Norton, 1978.

NOTES

29 K. E. Boulding, 'The economics of the coming space-ship earth', in H. Jarrett (ed.) *Environmental Quality in a Growing Economy*, Baltimore: Johns Hopkins Press for RFF, 1966, 3–14.

30 K. E. Boulding, *Ecodynamics: A New Theory of Societal Evolution*, Beverly Hills and London: Sage Publications; *idem, Evolutionary Economics*, Beverly Hills and London: Sage Publications, 1981.

31 P. Nijkamp, 'Economic and ecological models: a qualitative multidimensional view', in A-M. Jansson, *op. cit.*, 1984, pp. 167–84.

32 E. F. Schumacher, *Small is Beautiful: Economics as if People Mattered*, New York: Harper and Row/London: Blond and Briggs, 1973.

33 It sounds like an old Latvian proverb: actually I made it up.

34 A. Harris, 'Radical economies and natural resources', *Int. J. Env. Studies* 21, 1983, 45–53.

35 E. B. Barbier, *Economics, Natural Resource Scarcity and Development. Conventional and Alternative Views*, London: Earthscan, 1989.

36 R. B. Norgaard, 'Environmental economics: an evolutionary critique and a plea for pluralism', *J. Env. Econ. and Mgmt* 12, 1985, 389–94. The notion of biophysical carrying capacity has been at the heart of the 'Limits to Growth' scenarios of the Club of Rome and its associates, first in the 1970s but recently revived again by that group (D. H. Meadows, D. L. Meadows and J. Randers, *Beyond the Limits: Global Collapse or a Sustainable Future?*, London: Earthscan, 1992) and in more popular form by R. Douthwaite, *The Growth Illusion*, Bideford, Devon: Green Books, 1992.

37 W. Leiss, *The Limits to Satisfaction: an Essay on the Problem of Needs and Commodities*, Toronto and Buffalo: University of Toronto Press, 1976.

38 K. E. Boulding, 'What went wrong with economics?', *American Scientist* 30, 1986, 5–12.

39 E. J. Mishan, *The Economic Growth Debate: An Assessment*, London: Allen and Unwin, 1977. For an examination of a new generation of economic indicators which are less oriented towards conventional economic growth, see V. Anderson, *Alternative Economic Indicators*, London, 1991.

40 K. E. Boulding, *op. cit.*, 1986.

41 Theory and practice together are the subject of J. P. Lester (ed.), *Environmental Politics and Policy: Theories and Evidence*, Durham, NC: Duke University Press, 1991.

42 M. Q. Sibley, 'The relevance of classical political theory for economy, technology and ecology', *Alternatives*, Winter 1973, 14–35.

43 J. Fernie and A. S. Pitkethly, *Resources: Environment and Policy*, London: Harper and Row, 1985.

44 R. J. Barnett, *The Lean Years: Politics in the Age of Scarcity*, New York: Simon and Schuster, 1980/London: Sphere Books, 1981.

45 P. Lowe and J. Goyder, *Environmental Groups in Politics*, London: Allen and Unwin, 1983.

46 A. Gorz, *Ecology as Politics*, Boston: South End Press, 1980. First published in France as *Ecologie et Politique*, Paris: Editions Galilee, 1975.

47 A Conservative MP in the UK Parliament wrote in *The Guardian* newspaper during 1991, 'we are all put under pressure by a new group of what I call eco-terrorists. Many of them are ex-socialists and their green wellies have red linings.'

48 See e.g. P. C. Gould, *Early Green Politics: Back to Nature, Back to the Land, and Socialism in Britain, 1880–1900*, Brighton: Harvester Press, 1989; S. P. Hays, *Beauty, Health and Permanence: Environmental Politics in the U.S., 1955–85*, Cambridge: Cambridge University Press, 1987. A wider cast is made in

179

discussing e.g. socialism and anarchy in R. Eckersley, *Environmentalism and Political Theory: Toward an Ecocentric Approach*, London: UCL Press, 1992.

49 W. Ophuls, *Ecology and the Politics of Scarcity: Prologue to a Political Theory of the Steady State*, San Francisco: Freeman, 1977.

50 B. Tokar, 'Exploring the new ecologies. Social ecology, deep ecology and the future of Green political thought', *Alternatives* 15, 1988, 31–43. There is a similar conclusion in J. Young, *Post Environmentalism*, London: Belhaven Press, 1990, p. 201: 'The best chances for survival are for those social states which will be reached in irregular fashion, as individuals and families, streets or villages, suburbs, cities and the more sensible national governments.'

51 See the review by A. Bramwell, 'Blending ecological belief with political action', *Times Higher Education Supplement*, 23 June 1989, 19, and her article, 'Widespread seeds of the Green revolution', *ibid.*, 20 November 1987, 13. Think also how conservative, racist and paternalistic small communities can be.

52 For a global institutional perspective see J. S. Dryzek, *Rational Ecology: Environment and Political Economy*, 1987. Geographer R. J. Johnston takes the thesis that it is competitive capitalism that causes the degradation of environment and shows how this works out at local, national and global scales: R. J. Johnston, *Environmental Problems: Nature, Economy and State*, London: Belhaven, 1989. In this he echoes S. Schnaiberg, *vide* note 100.

53 K. E. Boulding, *Ecodynamics: A New Theory of Societal Evolution*, Beverly Hills and London: Sage, 1978.

54 The term and the idea are from W. Ophuls, *op. cit.* For a history of environmental, social and political attitudes and movements, see J. McCormick, *The Global Environmental Movement*, London: Belhaven Press, 1989.

55 A short bibliography would include the books by A. Dobson, *Green Political Thought: An Introduction*, London: Routledge, 1990; A. Atkinson, *Principles of Political Ecology*, London: Belhaven Press, 1991; R. L. Bryant, 'Putting politics first: the political ecology of sustainable development', *Global Ecology and Biogeography Letters* 1, 1991, 164–6. The means rather than the ends are stressed in D. Wall, *Getting There: Steps to a Green Society*, London: Green Print, 1990.

56 See especially chapter 7 of Yrjö Haila and R. Levin, *Humanity and Nature: Ecology, Science and Society*, London: Pluto Press, 1992.

57 This section relies heavily on L. Winner, *The Whale and the Reactor: A Search for Limits in an Age of High Technology*, Chicago and London: University of Chicago Press, 1986, I.1.

58 M. Bookchin, *The Ecology of Freedom: The Emergence and Dissolution of Hierarchy*, Palo Acto, CA: Cheshire Books, 1982.

59 His main synthesis is *Catastrophe or Cornucopia: The Environment, Politics and the Future*, Chichester: Wiley, 1982. Papers by S. F. Cotgrove of considerable interest, with more detail for the UK in the 1970s, include 'Technology, rationality and domination', *Social Studies of Science* 5, 1975, 55–78; (with A. Duff), 'Environmentalism, middle class radicalism and politics', *Soc. Rev.* 28, 1980, 333–51; *idem*, 'Environmentalism, values and social change', *Brit. J. Sociol.* 32, 1981, 92–110.

60 The founding document was D. H. Meadows *et al.*, *The Limits to Growth*, London: Earth Island Press, 1972. An extensive critique is in H. S. D. Cole *et al.*, *Thinking about the Future: A Critique of the Limits to Growth*, London: Chatto and Windus, 1973.

61 F. Hirsch, *Social Limits to Growth*, London: RKP, 1977. His arguments are much more complex than can adequately be summarised here.

62 T. Roszak, *Where the Wasteland Ends: Politics and Transcendence in Postindustrial Society*, Garden City, NY: Anchor Books, 1973.

63 T. Roszak, *Person/Planet: The Creative Disintegration of Industrial Society*, Garden City, NY: Anchor Press, 1978/London: Granada Books, 1981.

64 The relevant books by Marcuse were all published simultaneously by Beacon Hill in Boston and Allen Lane in London. Especially relevant here is *Counterrevolution and Revolt*, 1972, but see also the 'Bible' of the drop-out and alternative movements of the late 1960s, *One-Dimensional Man: Studies in the Ideology of Advanced Industrial Society*, 1964.

65 M. Bookchin, *The Ecology of Freedom: The Emergence and Dissolution of Hierarchy*, Palo Alto: Cheshire Books, 1982.

66 See J. Galtung, 'The Green movement: a socio-historical explanation', in M. Albrow and E. King (eds) *Globalization, Knowledge and Society*, London: Sage, 1990, pp. 235–50.

67 J. Porritt, *Seeing Green: The Politics of Ecology Explained*, Oxford: Blackwell, 1984; C. Spretnak and F. Capra, *Green Politics: The Global Promise*, New York: Dutton, 1984/London: Paladin Books, 1986. Porritt's book is very strong on the reasons for having Green politics, whereas the other volume devotes a great deal of its space to Green politics in West Germany, as well as having some discussion of the movement worldwide. On social and economic changes, see D. Wall, *Getting There: Steps to a Green Society*, London: Green Print, 1990.

68 The interaction of geography and philosophy is dealt with by R. J. Johnston, *Philosophy and Human Geography: An Introduction to Contemporary Approaches*, London: Edward Arnold, 1986, 2nd edn.

69 See R. J. Johnston, *Geography and Geographers: Anglo-American Human Geography since 1945*, London: Edward Arnold, 1991, 4th edn. There is a concise criticism of both positivist science and phenomenological approaches to Geography in N. Smith, 'Geography, science and post-positivist modes of explanation', *Progress in Human Geography* 3, 1979, 356–83.

70 This work is accessibly described in M. Chisholm, *Human Geography: Evolution or Revolution?* Harmondsworth: Pelican Books, 1975; and in R. J. Johnston *et al.* (eds), *The Dictionary of Human Geography*, Oxford: Blackwell, 1986, 2nd edn.

71 His book *Locational Analysis in Human Geography* (London: Edward Arnold, 1965) is something of a founding document and has been influential in other disciplines, such as archaeology.

72 M. Foucault, *Power/Knowledge*, Brighton: Harvester Press, 1980, ch. 4: 'Questions on Geography'.

73 There were exceptions. See e.g. W. Zelinsky, 'Beyond the exponentials: the role of geography in the great transition', *Geographical Review* 46, 499–535; S. R. Eyre, *The Real Wealth of Nations*, London: Edward Arnold, 1978.

74 D. Harvey, 'Population, resources and the ideology of science', *Economic Geography* 50, 1974, 256–77.

75 See ch. 5 of M. Quaini, *Geography and Marxism*, Oxford: Blackwell, 1982 (first publ. in Italian in 1974) (trans A. Braley, ed. R. King). Other treatments of Marxism in Geography can be found in e.g. D. Harvey, 'On the history and present condition of geography: an historical materialist manifesto', *Professional Geographer* 36, 1984, 1–11, and in his entry on 'Marxist geography' in R. J. Johnston *et al.* (eds) *op. cit.*, pp. 287–92. M. Eliot Hurst, 'Geography has neither Existence nor Future', in R. J. Johnston (ed.) *The Future of Geography*, London and New York: Methuen, 1985, pp. 59–91, takes the Marxist view of social science to the point of demanding a total reconstruction of it: 'to radically criticize, and deconstruct if need be, any current practice' (p. 84).

76 Advocates of this point of view suggest that it can be expressed clearly if unacademically as, 'Pull up the ladder, Jack, I'm on board'. More seriously, see the discussion by M. Fitzsimmons, 'The matter of nature', *Antipode* **21**, 1989, 106–20.

77 See S. R. Eyre and G. R. J. Jones (eds), *Geography as Human Ecology*, London: Edward Arnold, 1967.

78 A renewed case for this integration is made by R. W. Kates, 'The human environment: the road not taken, the road still beckoning', *Ann. Assoc. Amer. Geogr.* **77**, 1987, 525–34.

79 See my 'No rush to grow green', *Area* **22**, 1991, 384–7.

80 Examples of the first include R. Dyson-Hudson and E. A. Smith, 'Human territoriality, an ecological reassessment', *American Anthropologist* **80**, 1978, 21–41; P. Burnham and R. F. Ellen (eds), *Social and Ecological Systems*, London: Academic Press, 1979; R. F. Ellen, *Environment, Subsistence and System*, Cambridge University Press, 1982. On the second, see R. Layton (ed.), *Uluru, an Aborginal History of Ayers Rock*, Sydney: Australian Institute for Aboriginal Studies, 1986; *idem* (ed.), *Who Needs the Past?*, London: Unwin Hyman, 1989. P. Richards ('Saving the rain forest? – contested futures in conservation', in S. Wallmar (ed.) *The Anthropology of the Future*, London: Routledge, forthcoming) talks of the inhabitants of the Gola forest in Sierra Leone as regarding themselves as being under the protection of the forest. This is in contrast to nearly all the 'conservationist' views of tropical forests, who stand over the forest in order to protect it.

81 B. S. Orlove, 'Ecological anthropology', *Ann. Rev. Anthropol.* **9**, 1980, 235–73.

82 See, for example, S. Cook, 'Production ecology and economic anthropology: notes toward an integrated frame of reference', *Social Science Information* **12**, 1973, 25–52.

83 E. Boserup, *The Conditions of Agricultural Growth; the Economics of Agrarian Change Under Population Pressure*, London: Allen and Unwin, 1965.

84 See for example J. W. Bennett, *The Ecological Transition: Cultural Anthropology and Human Adaptation*, Oxford: Pergamon Press, 1976. Also, D. L. Hardesty, *Ecological Anthropology*, New York: Wiley, 1977.

85 The work of P. Richards is especially interesting in this context; see his *Indigenous Agricultural Revolution: Ecology and Food Production in West Africa*, London: Allen and Unwin, 1986.

86 R. F. Ellen, *Environment, Subsistence and System: The Ecology of Small-Scale Social Formations*, Cambridge: Cambridge University Press, 1982.

87 We might note with interest the publication in 1991 of a double number (118–19) of *Etudes Rurales* entitled 'La météo. Pour une anthropologie du temps qu'il fait'.

88 C. Geertz, *Works and Lives: the Anthropologist as Author*, Cambridge: Polity Press, 1988.

89 To this is added the tension created by interventions such as Environmental Impact Analysis.

90 See L. Allison, *Environmental Planning: A Political and Philosophical Analysis*, London: Allen and Unwin, 1974. Jeremy Bentham (1748–1832) was an English philosopher and jurist whose formulation of the idea of the greatest good for the greatest number is usually labelled utilitarianism. See also M. Breheny and A. Hooper (eds), *Rationality in Planning*, London: Pion 1985.

91 There is a comprehensive discussion in A. Paludi, *Planning Theory*, Oxford: Pergamon Press Urban and Regional Planning Series, vol. 7, 1973. See also Parts

I and V of A. Paludi (ed.), *A Reader in Planning Theory*, Oxford: Pergamon Press, Urban and Regional Planning Series vol. 5, 1973.

92 D. C. Eversley, *The Planner in Society: The Changing Role of a Profession*, London: Faber and Faber, 1973. The book is nearly all about the UK but the general lessons seem to apply universally. It is interesting to watch planning authorities come to terms with 'Green' thinking and see the spectrum of response from planting a few more trees to quite radical ideas about energy use and transport, for instance.

93 T. C. Daniel and J. Vining, 'Methodological issues in the assessment of landscape quality', in I. Altman and J. Wohwill (eds) *Behaviour and the Natural Environment*, New York and London: Plenum Press, Human Behavior and Environment. Advances in Theory and Research vol. 6, 1983, pp. 39–84.

94 Two reviews of the actual process in the UK are: D. G. Robinson *et al.*, *Landscape Evaluation*, University of Manchester (UK) Centre for Urban and Regional Research, 1976; E. C. Penning-Rowsell *et al.*, *Public Evaluation of Landscape Quality*, Enfield: Middlesex Polytechnic Planning Research Group, 1977. For the USA see E. H. Zube *et al.*, *Landscape Assessment: Value Perceptions and Resources*, Stroudsburg, PA: Dowden, Hutchinson and Ross, 1975. There is a review paper by E. C. Penning-Rowsell, 'Fluctuating fortunes in gauging landscape value', *Progress in Human Geography* 5, 1981, 25–51.

95 J. Appleton, *The Experience of Landscape*, London: Wiley, 1975. See also his 'Landscape evaluation: the theoretical vacuum', *Trans. Inst. Brit. Geogr.* 66, 1975, 120–3, and *The Symbolism of Habitat: an Interpretation of Landscape in the Arts*, Seattle, WA: University of Washington Press, 1990.

96 Note the attempt to formulate a theory in the tradition of the natural sciences.

97 This section depends entirely upon the work of the geographer Gilbert White and his students. The immediate sources are G. F. White (ed.), *Natural Hazards: Local, National, Global*, New York: Oxford University Press, 1974; I. Burton, R. W. Kates and G. F. White, *The Environment as Hazard*, New York: Oxford University Press, 1978; R. W. Kates and I. Burton (eds), *Geography, Resources and Environment*, Vol. II. *Themes from the Work of Gilbert F. White*, Chicago and London: University of Chicago Press, 1986, especially the essays by H. Kunreuther and P. Slovic, 'Decision-making in hazard and resource management', pp. 153–87, A. V. T. Whyte, 'From hazard perception to human ecology', pp. 240–71, T. O'Riordan, 'Coping with environmental hazards', pp. 272–309.

98 H. A. Simon, 'Theories of decision-making in economics and behavioral science', *American Economic Review* 49, 1959, 253–83.

99 K. Hewitt, 'The idea of calamity in a technocratic age', in K. Hewitt (ed.) *Interpretations of Calamity*, London: Allen and Unwin, 1983, Risks and Hazards Series no. 1, 3–32. See also E. Waddell, 'The hazards of scientism: a review article', *Human Ecology* 5, 1977, 69–76.

100 This paragraph follows closely the arguments in A. Schnaiberg, *The Environment from Surplus to Scarcity*, New York and Oxford: Oxford University Press, 1980.

101 There is a discussion of the role of social science in environmentalist thought in S. Yearley, *The Green Case: A Sociology of Environmental Issues, Arguments and Politics*, London: Harper Collins Academic, 1991: 'Afterword. Social science and the Green case', pp. 184–6. He claims that social science has theoretical perspectives that are crucial to an understanding of the whole issue. Examples are in the relations of environment, development and justice; and in the epistemological uncertainties of the natural sciences. The discussion is largely confined to the examination of the Green viewpoint itself.

102 Winner (see ch. 2) argues that certain technologies require specific ways of organising power and authority in society. He gives us this example of an enterprise at a large scale, namely that of the development of railways.

103 This is the theme of B. Fay, *Social Theory and Political Practice*, London: Allen and Unwin, 1975.

104 H. Newby, 'Join forces in modern marriage', *The Times Higher Education Supplement*, 17 January 1992, 20.

105 Another one would be strongly influenced by the deconstructive school of postmodernism. Consider these quotations (reproduced minus the Harvard-style citations in them) from P. M. Rosenau, *Post-modernism and the Social Sciences: Insights, Inroads, and Intrusions*, Princeton, NJ: Princeton University Press, 1992, 75–6, 'Charting any space (social, physical or cognitive) raises questions about the adequacy of representation . . . there are no true maps. . . . Conventional history collapses without subjects. Post-modernists reject Enlightenment science, a view consistent with a radical post-modern redefinition of linear time and its re-conceptualization of space or conventional geography.'

4 THE LIFEWORLD

1 D. C. D. Pocock, *The Nature of Environmental Perception*, University of Durham, Department of Geography Occasional Publication (New Series) no. 4, 1974.

2 D. C. D. Pocock, *op. cit.*

3 See, for example, R. G. Golledge and G. Zannaras, 'Cognitive approaches to the analysis of human spatial behaviour', in W. H. Ittelson (ed.) *Environment and Cognition*, New York and London: Seminar Press, 1973, pp. 59–94.

4 Yi Fu Tuan, *Topophilia. A Study of Environmental Perception, Attitudes, and Values*, Englewood Cliff, NJ: Prentice-Hall, 1974.

5 Venice is a good example: compare any photograph of it in a travel brochure or tourist guide with the number of people likely to be there when you are.

6 See e.g. T. Hägerstrand, 'Presence and absence: a look at conceptual choices and bodily necessities', *Regional Studies* 18, 1984, 373–80; T. Carlstein, *Time Resources, Society and Ecology*, vol. 1, *Preindustrial Societies*, London and Boston: Allen and Unwin, 1982.

7 My own example would be the difference between writing, as now when each look at my watch shows that at least an hour has passed since I last looked, and a long plane journey, when never more than 10 minutes have elapsed between glances.

8 Not to mention those who live in the past. This latter seems to be institutionalised in the UK now as 'Heritage' and indeed is a major growth industry in regions missing out on 'sunrise' industries and service employment. It all seems like a flight from an uncertain future into a past where there is at least some certainty (if only by virtue of the incomplete selection of 'facts'); this way also lies authoritarianism, to my mind.

9 Commonly quoted authors are H. Spiegelberg, *Doing Phenomenology*, The Hague: Nijhoff, 1975; A. Schutz, *The Phenomenology of the Social World*, London: Heinemann, 1972 (first publ. in German in 1932). Also W. Dilthey, *Gesammelte Schriften 1914–74* in 17 vols; for a shorter lead-in see C. Rose, 'Wilhelm Dilthey's philosophy of historical understanding. A neglected heritage of contemporary humanistic geography', in D. R. Stoddart (ed.) *Geography, Ideology and Social Concern*, Oxford: Blackwell, 1981, pp. 99–133. An interesting, though rather impressionistic, study of how the 'self' changes in its perception

and cognition of the world through time and how it is conditioned by culture is Yi Fu Tuan, *Segmented Worlds and Self*, Minneapolis: University of Minnesota Press, 1982.

10 Some writers compare this to the idea of everybody (and every living thing perhaps) going around in a bubble that constitutes their lifeworld. The symmetry of a bubble is perhaps misleading.

11 Like James Joyce, the geographer and planner Gunnar Olsson makes a virtue of linguistic ambiguity. See his 'Of ambiguity or far cries from a memorializing mamafesta', in D. Ley and M. S. Samuels (eds) *Humanistic Geography. Prospects and Problems*, London: Croom Helm, 1978, pp. 109–20.

12 D. Seaman, 'The phenomenological contribution to environmental psychology', *J. Env. Psychology* 2, 1982, 119–40. A very large-scale treatment of the application of phenomenological thought to the conventional divisions of Euro-American social science is M. Natanson (ed.), *Phenomenology and the Social Sciences*, Evanston, IL: Northwestern University Press, 1973, 2 vols, North-western Studies in Phenomenology and Existential Philosophy. There is a very wide-ranging 'Introduction' by the editor (3–44), and a chapter or chapters on Anthropology, Sociology, Psychology, Linguistics, History, Political Science, Economics and Legal Theory.

13 A clear introduction is given in M. Pusey, *Jürgen Habermas*, Chichester: Ellis Horwood, Key Sociologists Series, 1987. A selection of original works and commentary putting him and others into context is K. Mueller-Vollmer (ed.), *The Hermeneutics Reader*, New York: Continuum, 1989.

14 N. Everden, *The Natural Alien: Humankind and Environment*, Toronto, Buffalo and London: University of Toronto Press, 1985. He uses the label 'Environmentalist' rather than Catastrophist but the concept is much the same. Everden's book contains extended introductions to the writings of Merleau-Ponty and Heidegger which serve as very useful guides for beginners like me. Heidegger in particular seems very hard to understand, and I am not sure I have done justice to his thought. It seems to be generally acknowledged that his writing is somewhat opaque.

15 This is capitalised to indicate that is of the nature of a technical term; it has nothing to do with Supreme Beings.

16 The title of Peter Sellers' last movie. I wonder . . .

17 There is a resonance here of John Donne's famous 'No man is an island . . .' On the other hand, his also much-quoted *An Anatomy of the World* suggests the opposite if the cosmos and human relations could be seen to be isomeric: ''Tis all in pieces, all coherence gone;/All just supply and all relation'.

18 D. Abram, 'Merleau-Ponty and the voice of the Earth', *Environmental Ethics* 10, 1988, 101–20.

19 D. Abram, 'The perceptual implications of Gaia', *The Ecologist* 15, 96–103.

20 An introduction is provided by W. J. Mills, 'Positivism reversed: the relevance of Giambattista Vico', *Trans. Inst. Brit. Geogr.* (NS) 7, 1982, 1–14. The main edition of Vico's *New Science* is from 1731; a modern version is translated by T. G. Bergin and M. H. Fisch, *The New Science of Giambattista Vico*, Ithaca, NY: Cornell University Press, 1968. See also L. Pompa, *Vico: a Study of the New Science*, Cambridge: Cambridge University Press, 1975. Vico was sure that imaginative insight was always necessary, however, to understand other human societies, especially those of the past. He called that insight *Fantasia*. See I. Berlin, 'Giambattista Vico and cultural history', in H. Hardy (ed.) *The Crooked Timber of Humanity: Chapters in the History of Ideas*, London: Fontana, 1991, pp. 49–69.

21 Vico thought of the totality of human history as cyclic.

22 In general, aesthetics deal with the immediate affective qualities of the contemplation of a human production. But the criteria may also be applied to nature, for example, or the human body. Thus a taste in painting style may create a set of values for judging actual landscapes: hence probably the use of the term 'picturesque' to describe an outdoor scene.

23 For instance, W. H. Auden (*Letter to Lord Byron, III*):

> To me art's subject is the human clay,
> And landscape but a background to a torso;
> All Cézanne's apples I would give away
> For one small Goya or a Daumier.

I think there is evidence elsewhere in his poetry to suggest that there was more to his involvement with the land than that.

24 From J. Hospers, *Understanding the Arts*, Englewood Cliffs, NJ: Prentice-Hall, 1982, though I doubt that he would claim it as original.

25 And yet . . . this seems incomplete in the light of statements like Georges Braque's 'Art is meant to disturb. Science reassures'. But if we embark on a discussion of 'why and what for' then there will be no space to discuss any actual arts themselves.

26 The list ought to include music, TV, cinema, opera, garden design, video, architecture, radio and other sound-pictures.

27 See K. Clark, *Landscape into Art*, London: John Murray, 1949, and later editions; P. Shepard, *Man in the Landscape: A Historic View of the Esthetics of Nature*, New York: Knopf, 1976; E. Carli, *The Landscape in Art*, New York: William Morrow, 1980, first published in Italy in 1979.

28 K. Clark, *op. cit.*, 1949.

29 This relationship can be symbolised by the depiction of animals for example rather than the whole landscape. See Kenneth Clark, *Animals and Man*, London: Thames and Hudson, 1977. The use of details from larger canvases sometimes rather vitiates to my mind some of the interpretations that he offers.

30 For obvious reasons, this account is limited by what I have read (as are many of the others) and what I have seen in both reproductions and the originals, but above all by what has impressed me, which may well be a far cry from received opinion. There is no substitute for seeing for oneself, and especially the originals even given the excellence of modern colour printing.

31 For Bellini, *The Madonna of the Meadow* exemplifies these traits; for Piero di Cosimo, *The Death of Procris* and *Forest Fire*. For a general and iconographic view, see A. R. Turner, *The Vision of Landscape in Renaissance Italy*, Princeton University Press, 1966 (paperback edn 1974).

32 The chronology of his paintings seems obscure; he lived from 1477 or 1478 to 1510, mostly in Venice. See B. Wittkower, 'Georgione and Arcady', in R. W. Wittkower (ed.) *Idea and Image. Studies in the Italian Renaissance*, London: Thames and Hudson, 1978.

33 I spent much of my childhood in east Lincolnshire and can testify to the enormous importance of the sky in the landscape. And until I was 11 or 12 I never realised that streams could actually have moving water in them. More to the point, see W. Stechow, *Dutch Landscape Painting of the Seventeenth Century*, New York: Hacker, 1980.

34 As no doubt they were meant to be, especially the religious paintings which presumably tried to provoke awe or guilt; I am sceptical that the many nativities evoked much feeling of affection, however.

35 It seems that the Dutch word *landskip* which referred to a genre of paintings

('hills, woods, castels, seas, ruins, valleys, hanging rocks, cities, towns', wrote Henry Peacham in 1606 of this type of art) was responsible for the introduction of the word *landscape* into English. The S.O.D. gives 1632 for its first usage and also notes that the term was used in opposition to seascape.

36 P. Shepard, *op. cit.*, p. 125. Constable's paintings are put into an interpretive context along with e.g. Gainsborough in A. Bermingham, *Landscape and Ideology: The English Rustic Tradition 1740–1860*, Berkeley and Los Angeles: University of California Press, 1986.

37 Turner and Wordsworth are compared in J. A. W. Heffernan, 'Reflections on reflections in English Romantic poetry and painting', in H. R. Garvin (ed.) *The Arts and their Interrelations*, Lewisburg, PA: Bucknell University Press/ London: Associated University Presses, 1978 (vol. **24** no. 2 of *Bucknell Review*). This adds credence to the idea of Turner witnessing the passing of the pre-industrial age, as did Wordsworth when for example he fulminated against the coming of the railway to Coniston. For a wider discussion of landscape as a cultural expression see D. Cosgrove and S. Daniels (eds), *The Iconography of Landscape*, Cambridge: Cambridge University Press, 1988.

38 These men were most active in the period 1863–86, which covers the period of the seven Impressionist exhibitions in Paris. Gauguin went to Tahiti in 1895.

39 The early Monets in the Orsay are supplemented by the collection in the Musée Marmottan in Paris; I have found very helpful J. House, *Monet: Nature into Art*, Yale University Press, 1986.

40 A content analysis of English painting in the last 200 years shows that between 1790 and 1830 industry was often selected as a subject: grist mills, lime kilns, quarries and canals were often singled out. After 1830, industrial scenes were rare and even more so after 1850; railways were common until 1850. Recovery of heavy industry as a subject took place after 1910 but often in the context of its decline. See P. Howard, *Landscape: The Artists' Vision*, London: Routledge, 1991.

41 D. Cosgrove and J. E. Thornes, 'Of truth of clouds: John Ruskin and the moral order in landscape', in D. C. D. Pocock (ed.) *Humanistic Geography and Literature: Essays on the Experience of Place*, London, Croom Helm, 1981, pp. 20–46.

42 R. Cafritz, L. Gowing and D. Rosand, *Places of Delight: The Pastoral Landscape*, Washington, DC: The Phillips Collection, 1988.

43 H. Read, *A Concise History of Modern Sculpture*, London: Thames and Hudson, 1974.

44 This point was strongly made in the UK's Channel Four TV series *A Love Affair with Nature*, first broadcast in 1987.

45 See for example Jean Dubuffet's *Natura Genetrix* (1952) which shows the earth as a kind of compost: a coagulation of shapes and textures.

46 His 1991 exhibition in the Hayward Gallery, London, was called 'Walking in Circles'.

47 With regret: I should think they have a lot to tell us about popular cognition of the environment. Look at the way the Japanese always include people in their shots, for example. 'Naive' photography can be seen as high art, witness the published work of Jimmy Forsyth, who recorded with an old folding camera the disappearing terraces of nineteenth-century housing and their life in the West End of Tyneside after 1950 (J. Forsyth, *Scotswood Road*, Newcastle-upon-Tyne: Bloodaxe Books, 1986). The essence of the factory method of production is reflected in the colourless sameness of the pictures. It is now commonplace to romanticise this type of community.

48 There are numerous volumes of his work. An introductory selection is in his *Yosemite Valley*, San Francisco, CA: 5 Associates, 1959.

49 F. Goodwin and T. Hughes, *Remains of Elmet*, London: Faber and Faber, 1979. This is a powerful antidote to the 'Country-house' type of landscape photography.

50 The 'quality' newspapers occasionally have an art-form picture with only a simple caption.

51 The best electrical systems, we may remember, have an Earth wire.

52 The poet A. E. Housman talked of the test of true poetry being that it raised the hairs on the back of his neck while shaving. We need an equivalent for women. Geoffrey Grigson (ed.), *The Faber Book of Poems and Places*, London: Faber and Faber, 1980 seems to me a good example of a lot of scenery rather than a lot of real poetry.

53 For example, see Annette Kolodny, *The Lay of the Land: Metaphor as Experience and History in American Life and Letters*, Chapel Hill, NC: University of North Carolina Press, 1975; R. Lawson-Peebles, *Landscape and Written Expression in Revolutionary America: The World Turned Upside Down!*, Cambridge: Cambridge University Press, 1990. The carrying-up of various sorts of writings into a unified story of how the land was decisive in forming the ideologies of the European settlers is treated in M. Jehlen, *American Incarnation: The Individual, the Nation and the Continent*, Cambridge, MA and London: Harvard University Press, 1986. It can also be argued that the Europeans had to depict the continent as a wilderness to justify evicting, by one means or another, its aboriginal human inhabitants.

54 D. C. D. Pocock (ed.), *Humanistic Geography and Literature: Essays on the Experience of Place*, London: Croom Helm, 1981. See also R. Paulson, *Literary Landscapes: Turner and Constable*, New Haven: Yale University Press, 1982.

55 M. Drabble, *A Writer's Britain: Landscape in Literature*, London: Thames and Hudson, 1979.

56 Just as there is no good English equivalent of the German *Weltanschauung*, so the word *Zeitgeist* does not translate easily: the equivalent employed here is comparatively clumsy.

57 The Roskill Report on the siting of a new airport for London (1971) included a CBA which valued a very old church (as I remember, some of it was Saxon) simply at its fire insurance level. But what other figure could have been included? (That site was not chosen.)

58 Translated by R. Mannheim, Picador Books, London, 1988.

59 A number of characters from earlier Grass novels appear also, notably from *Tin Drum* and *The Flounder*. The latter can be recommended ('A celebration of life, food and sex' says the cover of the Penguin edition of 1979, but the title page is innocent of anything but the eponymous flatfish) as an antidote to the rather pessimistic outlooks of both *Tin Drum* (of which there is an excellent movie even though it deals with only half the story) and *The Rat*.

60 I use the Japanese custom of putting the family name first. Kawabata lived from 1899–1972.

61 Translation by E. G. Seidensticker. An eight-mat room is about 4 × 4 metres.

62 An allusion to the rise of awareness once the self's physical nature has been 'cleared'. From the medieval Kannada of Mysore: see A. K. Ramanujan, *Speaking of Siva*, Harmondsworth: Penguin, 1973, p. 160.

63 His prose work *Guide to the Lakes* (1822) carried through his self-avowed calling as poet, art critic and landscape gardener. In the manner of a letter-writer today to the quality newspapers, he was against landscape changes such as the coming of the railway to Coniston, the planting of larches and the whitewashing of cottages. We would give him only 30 per cent for such judgements, I suppose.

64 J. Bate, *Romantic Ecology: Wordsworth and the Environmental Tradition*, London and New York: Routledge, 1991.
65 In *Song at the Year's Turning*, London: Rupert Hart-Davis, 1955, p. 54.
66 *Ibid.*, p. 112, 'A Welshman to any Tourist'.
67 In *High Windows*, London: Faber and Faber, 1974.
68 G. Smith, *The Waste Land*, London: Allen and Unwin, 1983.
69 I went there in the very hot summer of 1976 and it was indeed just like that, including the earlier lines on the deep lane 'shuttered with branches'.
70 D. Traversi, *T. S. Eliot: The Longer Poems*, London: Bodley Head, 1976.
71 F. I. Carpenter, *Robinson Jeffers*, New York: Twayne Publishers, 1962, ch. 5.
72 H. Y. Jung and P. Jung, 'Gary Snyder's ecopiety', *Environmental History Review* 14, 1990, 75–87.
73 J. W. Beach, *Obsessive Images: Symbolism in Poetry of the 1930s and 1940s*, Westport, CT: Greenwood Press, 1973, especially 'Words from Geography and Travel', pp. 104–207.
74 There are many editions of this. Examples are D. Britton, *A Haiku Journey*, Tokyo: Kodansha International, 1974; Nobuyuki Yuasa, *The Narrow Road to the Deep North*, Harmondsworth: Penguin, 1966.
75 Anthologised with several other modern poems in G. Bownas and A. Thwaite, *The Penguin Book of Japanese Verse*, Harmondsworth: Penguin, 1964.
76 The quotation is from C. Ives, *Memos* (ed. J Kirkpatrick), New York: Norton, 1972, p. 106 and quoted in J. N. Serio, 'The ultimate music is abstract: Charles Ives and Wallace Stevens', in H. R. Garvin (ed.) *The Arts and their Interrelations*, Lewisburg PA: Bucknell University Press/London: Associated University Presses, 1978 (*Bucknell Review* 24 (2), 1978), 120–31.
77 The eventual idea was to have 6 to 10 different orchestras placed on separate mountain tops, each moving in its own independent time orbit, and meeting each other when their time cycles ran out. Unhappily, only fragments of the piece were completed (Kirkpatrick, *op. cit.*, p. 108).
78 There is a parallel in the music of Harry Partch (1901–74) who lived among native Americans in the Western deserts but also lived the life of a hobo. His music aimed to use instruments made from local materials, intersperse native melodies with semi-articulate words and allusions to myth, and to rediscover 'the world and the body' as one. See T. McGeary (ed.), *Bitter Music*, Champaign, IL: University of Illinois Press, 1991.
79 R. Murray Schaefer, *The Tuning of the World*, New York: Knopf, 1977.
80 What would the course of WWI have been if napalm had become available to one side in say 1916?
81 Derived from *archi*tecture and *eco*logy. See J. Grange, 'The arcology of Paolo Soleri: technology as cosmology', in H. R. Garvin, *op. cit.*, 1978, pp. 170–87.
82 It is argued vigorously with examples from poetry and painting in two articles on 'The arts and planetary survival' by D. Trussell: (Part I), *The Ecologist* 19, 1989, 179–5; (Part II), *The Ecologist* 20, 1990, 4–8. What is not explored by me in the present work is whether there are features common to all the fine arts which (a) underlie their aesthetic appeal and (b) link them to the ecological world. I suspect that the relationship between order and variety might be an interesting avenue: it would be a characteristic of both paintings and ecosystems, for example. There are hints of this in J. F. Wohlwill, 'The place of order and uncertainty in art and environmental aesthetics', *Motivation and Emotion* 4, 1980, 133–42.
83 Readings of these movies in a broader context are given by D. Harvey, *The Condition of Postmodernity*, Oxford: Blackwell, 1990.

5 NORMATIVE BEHAVIOUR

1 There is an introductory bibliography by E. Katz, 'Ethics and philosophy of the environment: a brief review of the major literature', *Environmental History Review* 15, 1991, 79–86.

2 It was clearly necessary at one stage to establish that environment was a fit topic for ethical inquiry and to try to define the outer boundaries of that inquiry. See H. Rolston, 'Is there an ecological ethic?', *Ethics* 85, 1975, 93–109, reprinted in D. Scherer and T. Attig (eds) *Ethics and the Environment*, Englewood Cliffs, NJ: Prentice-Hall, 1983, pp. 41–53; T. Regan, 'The nature and possibility of an environmental ethic', *Environmental Ethics* 3, 1981, 19–34; C. A. M. Duncan, 'On identifying a sound environmental ethic in history: prolegomena to any future environmental history', *Environmental History Review* 15, 1991, 5–30. As an overview introduction to the field, I found K. S. Shrader-Frechette (ed.), *Environmental Ethics*, Pacific Grove, CA: The Boxwood Press, 1981, especially helpful.

3 **Ontology** deals with what can, or cannot, exist, although in the case of nature this can be taken as given. The ontology then focuses on this and the foundations of our behaviour in it.

4 E. C. Hargrove, *Foundations of Environmental Ethics*, Englewood Cliffs, NJ: Prentice Hall, 1989, esp pp. 191–205. The recent history of the philosophy–ethics continuum in this field (very largely in North America, it seems) is described by R. F. Nash in chapter 5 ('The greening of philosophy') of his book *The Rights of Nature*, Madison,WI and London: University of Wisconsin Press, 1989.

5 The context is given for the use of both the natural and social sciences by J. Petulla, 'Toward an environmental philosophy: in search of a methodology', *Environmental Review* 2, 1977, 14–43. See also the essays in J. R. Engel and J. G. Engel (eds), *Ethics of Environment and Development*, London: Belhaven Press, 1990, especially R. Kothari, 'Environment, technology, and ethics', pp. 27–35, and H. Rolston, 'Science-based versus traditional ethics', pp. 63–72.

6 B. S. Gower, 'What do we owe future generations?', in D. E. Cooper and J. Palmer (eds) *The Environment in Question: Ethics and Global Issues*, London: Routledge, 1992, pp. 1–12.

7 T. Regan, *op. cit.*, 1981.

8 J. Baird Callicott, 'Intrinsic value, quantum theory, and environmental ethics', *Environmental Ethics* 7, 1985, 257–75. Not for the faint-hearted reader.

9 See F. Mathews, *The Ecological Self*, London: Routledge, 1991.

10 See the examples of P. R. Hay, 'The contemporary environment movement as Neo-Romanticism: a re-appraisal from Tasmania', *Environmental Review* 12, 1988, 39–59. Also the extended treatment in A. Bramwell, *Ecology in the 20th Century: A History*, New Haven and London: Yale University Press, 1989.

11 M. Zimmerman, 'The critique of natural rights and the search for a non-anthropocentric basis for moral behaviour', *J. Value Enquiry* 19, 1985, 43–53.

12 G. Hardin, *The Voyage of the Spaceship Beagle: Exploring New Ethics for Survival*, New York: Viking Books, 1972/Harmondsworth: Pelican Books, 1973; *idem*, 'Living on a lifeboat', *BioScience* 24, 1974, 561–68.

13 P. Ehrlich, *The Population Bomb*, New York: Ballantine, 1968.

14 For a summary of this material see K. S. Shrader-Frechette, 'Alternative ethics regarding the environment', in *idem, Environmental Ethics*, Pacific Grove, CA: The Boxwood Press, 1981, pp. 28–44.

15 Most of this part of the discussion derives from R. Goodin, 'Ethical principles for environmental protection', in R. Elliot and A. Gare (eds) *Environmental*

Philosophy, University Park and London: Pennsylvania State University Press, 1983, pp. 3–20.

16 These two paragraphs take their material from chapters 6–9 of R. Attfield, *The Ethics of Environmental Concern*, Oxford: Blackwell, 1983.

17 M. Midgley, *Beast and Man: The Roots of Human Nature*, Hassocks, Sussex: Harvester Press, 1979. See also R. Attfield, 'Attitudes to wildlife in the history of ideas', *Environmental History Review* 15, 1991, 71–8.

18 See the historical discussion in chapter 5 of J. Passmore, *Man's Responsibility for Nature*, London: Duckworth, 1980, 2nd edn. Changes in sensibilities for one country for one 300–year period are illuminatingly discussed in K. Thomas, *Man and the Natural World: Changing Attitudes in England 1500–1800*, London: Allen Lane, 1983.

19 M. Midgley, *Animals and Why They Matter: A Journey Around the Species Barrier*, Harmondsworth: Penguin Books, 1983.

20 In May 1988 I saw the movie *The Unbearable Lightness of Being* which is filled with tragedy of various kinds. But it was the death of the dog that caused most tears in the audience.

21 See chapters 8 and 9 of R. Attfield, *op. cit.*

22 That is the idea that ideas can influence the practical world and that they should be graded in value by their success at the pragmatic level.

23 See T. Regan and P. Singer (eds), *Animal Rights and Human Obligations*, Englewood Cliffs, NJ: Prentice-Hall, 1976; P. Singer, *Animal Liberation, A New Ethic for Our Treatment of Animals*, London: Cape, 1976; *idem, In Defense of Animals*, Oxford: Blackwell, 1985; S. R. L. Clark, *The Moral Status of Animals*, Oxford: Clarendon Press, 1977.

24 A defence of commensense attitudes towards animals is given by M. P. T. Leahy, *Against Liberation. Putting Animals into Perspective*, London and New York: Routledge, 1991.

25 C. Stone, *Should Trees Have Standing? Towards Legal Rights for Natural Objects*, Portola Valley, CA: Tioga Publishing Co, 1988, 2nd edn.

26 M. A. Warren, 'The rights of the nonhuman world', in Elliot and Gare, *op. cit.*, pp. 109–34. See also, B. G. Norton, *Why Preserve Natural Variety?*, Princeton, NJ: Princeton University Press, 1987.

27 cf. Alan Watts, 'The world is your body', in R. Disch (ed.) *The Ecological Conscience: Values for Survival*, Englewood Cliffs, NJ: Prentice-Hall, 1970, pp. 181–93.

28 The main source is A. Leopold, *A Sand County Almanac*, New York: Oxford University Press, 1949. Be prepared for surprises like the delight in hunting and shooting. The earlier formulation can be seen in e.g. 'The conservation ethic', in the *Journal of Forestry* for October 1933, reprinted in R. Disch, *The Ecological Conscience: Values for Survival*, Englewood Cliffs, NJ: Prentice-Hall, 1970, pp. 44–55.

29 Leopold starts his 1933 paper by remarking on the fact that there was no ethical barrier to Odysseus hanging a dozen slave-girls when he got home on suspicion of their misbehaviour.

30 Especially J. Passmore, *op. cit.*, 1980.

31 In several of his works, the distinguished humanist René Dubos was fond of using the eighteenth-century landscape gardens of England as an example of the beneficent effects of the human hand. I tried, by correspondence, to convince him that there was another side to this in the shape of the dispossession suffered by many smaller landholders but I don't think I was ever persuasive.

32 W. T. Blackstone, 'Ethics and ecology', in W. T. Blackstone (ed.) *Philosophy*

and Environmental Crisis, Athens, GA: University of Georgia Press, 1974, pp. 16–42.

33 R. Attfield, *op. cit.*, 1983, ch. 8.

34 J. R. Engel, 'Ethics', in D. C. Pitt (ed.), *The Future of the Environment: The Social Dimensions of Conservation and Ecological Alternatives*, London and New York: Routledge, 1988, pp. 46–59.

35 See the wide-ranging discussions in A. McLaughlin, 'Images and ethics of nature', *Environmental Ethics* 7, 1985, 239–319; and M. Sagoff, 'Fact and value in ecological science', *ibid.*, 99–118.

36 For the Gaia–human mind linkage see P. Russel, *The Awakening Earth: Our Next Evolutionary Leap*, London: RKP, 1982.

37 T. Roszak, *Person/Planet: The Creative Disintegration of Industrial Society*, New York: Doubleday/Anchor, 1978; London: Gollancz, 1979.

38 D. Bohm, *Wholeness and the Implicate Order*, London: RKP, 1982.

39 Although pointed out in Aldous Huxley's *The Perennial Philosophy* (London: Fontana, 1958), the most famous statement of it in the present context is F. Capra, *The Tao of Physics*, Berkeley, CA: Shambhala Press/London: Wildwood House, 1975.

40 Pierre Teilhard de Chardin, *The Phenomenon of Man*, London: Collins, 1965. For a primer and guide to Chardin, try J.-P. Demoulin (ed.), *Let Me Explain*, London: Collins, 1970. When typing this chapter into the word-processor I used 'de' as an abbreviation for Deep Ecology, then replacing it globally with the full spelling. The result in two or three places was Pierre Teilhard Deep Ecology Chardin, which might delight some people.

41 H. Skolimowsky, *Ecophilosophy: Designing New Tactics for Living*, London: Marion Boyars, 1979.

42 F. Capra, *The Turning Point: Science, Society and the Rising Culture*, New York: Simon and Schuster/London: Wildwood House, 1982.

43 R. Buckminster Fuller, *An Operating Manual for Spaceship Earth*, Carbondale, IL: Southern Illinois University Press, 1969.

44 It is customary to point out that the Chinese ideograph for 'crisis' contains two elements: *wei* for 'danger' and *chi* for 'opportunity'.

45 M. Eliade, *A History of Religious Ideas*, vol. 1. *From the Stone Age to the Eleusinian Mysteries*, Chicago and London: University of Chicago Press, 1982, trans. W. R. Trask. First published in French in 1978.

46 Eliade, *op. cit.*, vol. 2. *From Gautama Buddha to the Triumph of Christianity*.

47 M. Eliade, *The Myth of the Eternal Return or, Cosmos and History*, Princeton University Press, 1971 (original edition 1954), Bollingen Foundation Series XLVI.

48 A chronology of changes in religious (mostly Christian) thought in North America is given by R. F. Nash in chapter 4 ('The greening of religion') of his *The Rights of Nature: A History of Environmental Ethics*, Madison, WI and London: University of Wisconsin Press, 1989.

49 J. Kay, 'Human domination of nature in the Hebrew Bible', *Ann. Assoc. Amer. Geogr.* **79**, 1989, 214–32.

50 L. White, 'The historical roots of our ecologic crisis', *Science* **155**, 1967, 1203–7, and much anthologised since as well as heavily criticised by both Christians and those of secular views. A fine example of a really seminal paper, right or not.

51 Quoted by J. Cobb, *Ecology and Religion: Toward a New Christian Theology of Nature*, Ramsey, NJ: Paulist Press, 1983. This whole book is probably one of the most concise and coherent statements of both history and dogmatics available.

52 A. R. Peacocke, *Creation and the World of Science*, Oxford: Clarendon Press, 1979, The Bampton Lectures, 1978. A difficult book but remarkable in its comprehensiveness.

53 Lynn White (see n 50) proposed St Francis as the patron saint of ecologists. However, Francis is quoted as saying 'every creature proclaims "God made me for your sake, O man!"' (P. Singer, 'Not for humans only: the place of nonhumans in environmental issues', in K. E. Goodpaster and K. M. Sayre (eds) *Ethics and Problems of the 21st Century*, Notre Dame, IN and London: University of Notre Dame Press, 1979, pp. 191–206).

54 Inexplicably, this legend is omitted from the anthology edited by C. Bamford and W. Parker Marsh, *Celtic Christianity: Ecology and Holiness*, Edinburgh: Floris Books, 1982. In fact there is some nature in this book but not much ecology.

55 See, for example, M. Fox, *Illuminations of Hildegard of Bingen*, Sante Fé, NM: Bear and Co, 1985; G. Uhlein, *Meditations with Hildegard of Bingen*, Sante Fé, NM: Bear and Co, 1983. She was no mean musician, either: listen to the *Symphoniae* on CD Editio Classica GD 77020.

56 H. Schwarz, 'The eschatological dimension of ecology', *Zygon* **9**, 1974, 323–38.

57 R. H. Hiers, 'Ecology, biblical theology, and methodology: biblical perspectives on the environment', *Zygon* **19**, 1984, 43–59.

58 At its fullest expression in J. Black, *The Dominion of Man: The Search for Ecological Responsibility*, Edinburgh University Press, 1970. Subsequently subjected to strong criticism by J. Passmore, *Man's Responsibility for Nature*, London: Duckworth, 1980, 2nd edn, who argues that stewardship can only apply to the control of men by men. It is interesting to note that the Papal Encyclical of 1981 referred to on p. 130 constantly refers to the subduing aspect of Genesis but not at all to the stewardship or responsibility aspects *(Laborem exercens*, Encyclical Letter of the Supreme Pontiff John Paul II on Human Work, London: Catholic Truth Society, 1981).

59 P. Teilhard de Chardin, *op. cit.*

60 See S. McDonagh, *To Care for the Earth: A Call to a New Theology*, London: Chapman 1986; M. Fox, *Original Blessing*, Santa Fé, NM: Bear and Co, 1983; *idem, Creation Sprituality*, London: Harper Collins, 1991. Matthew Fox was silenced by the Vatican in the late 1980s. The fusing of environmentalism, feminism and religion can be seen in A. Primavesi, *From Apocalypse to Genesis*, London: Burns and Oates, 1991. Primavesi sees the fall not as a cosmic tragedy but as a coming to maturity.

61 See, for instance, J. V. Taylor, *Enough is Enough*, London: SCM Press, 1975, and the account of the low-impact lifestyle movement in H. Dammers, *Life Style: a Parable of Sharing*, Wellingborough, England: Turnstone Press, 1982.

62 A. R. Drengson, 'The sacred and the limits of the technological fix', *Zygon* **19**, 1984, 259–75. See also L. Gilkey, 'Nature, reality and the sacred: a meditation in science and religion', *Zygon* **24**, 1989, 283–98, and the extended discussion of 'sanctity' in H. Skolimowski, *Living Philosophy: Eco-philosophy as a Tree of Life*, London: Arkana Books, 1992.

63 J. B. Cobb, 'Ecology, science and religion: toward a postmodern worldview', in D. R. Griffin (ed.) *The Reenchantment of Science: Postmodern Proposals*, Albany, NY: SUNY Press, 1988, pp. 99–113.

64 J. Donald Hughes, *American Indian Ecology*, El Paso, TX: Texas Western Press, 1983; J. D. Hughes and J. Swan, 'How much of the earth is sacred space?', *Environmental Review* **10**, 1986, 247–59.

65 O. P. Dwivedi, B. N. Tiwari and R. N. Tripathi, 'The Hindu concept of ecology and the environmental crisis', *Indian J. of Public Administration* **30**, 1984, 33–67.

66 A well-known and somewhat flamboyant interpreter of the Tao and of Zen for westerners was Alan Watts. See his *Nature, Man and Woman*, New York: Vintage Books, 1970; first published New York: Pantheon Books, 1958.

67 D. E. Shaner, 'The Japanese experience of nature', in J. B. Callicott and R. T. Ames (eds) *Nature in Asian Traditions of Thought*, Albany, NY: SUNY Press, 1989, pp. 163–82. Another 'cross-cultural' comparison is made in D. E. Shaner and R. S. Duval, 'Conservation ethics and the Japanese intellectual tradition', *Environmental Ethics* 11, 1989, 197–214.

68 F. Katayama and M. Kurosaka (eds), *Resonance Between the Essence of Nature and the Human Mind*, Tokyo: Shisakusa Publishing Co., 1988, 3 vols (in Japanese).

69 F. H. Cook, 'The jewel net of Indra', in Callicott and Ames, *op. cit.*, pp. 213–29; *idem, Hua-yen Buddhism*, University Park, PA: Penn State Press, 1977.

70 I. H. Zaidi, 'On the ethics of man's interaction with the environment: an Islamic approach', *Environmental Ethics* 3, 1981, 35–47; IUCN, *Islamic Principles for the Conservation of the Environment*, Gland, Switzerland: IUCN, 1983.

71 R. Engel, 'Ethics', in D. C. Pitt (ed.) *The Future of the Environment: The Social Dimensions of Conservation and Ecological Alternatives*, London and New York: Routledge, 1988, pp. 23–45. In 1992, Cassel in London published five edited books constituting a set called *World Religions and Ecology*, with volumes on Buddhism, Hinduism, Christianity, Islam and Judaism.

72 The master statement is the book by A. Naess, *Ecology, Community and Lifestyle. Outline of an Ecosophy*, Cambridge: Cambridge University Press, 1989. Elsewhere, there appears to be a large and scattered literature. I have found the most accessible sources to be: R. C. Schultz and J. D. Hughes (eds), *Ecological Consciousness*, Washington, DC: University Press of America, 1981, especially the contributions by Dolores LaChapelle, pp. 295–324, and G. Sessions, pp. 391–463, which has a very large bibliography; B. Devall and G. Sessions (eds), *Deep Ecology: Living as if Nature Mattered*, Salt Lake City, UT: Peregrine Smith Books, 1985; G. Sessions, 'The Deep Ecology movement: a review', *Environmental Review* 11, 1987, 105–25; F. Matthews, 'Conservation and self-realization: a deep ecology perspective', *Environmental Ethics* 10, 1988, 347–55. Earlier work by Naess includes 'The deep ecology movement: some philosophical aspects', *Philosophical Inquiry* 8, 1986, 10–31, and 'The shallow and the deep, long-range ecology movements: a summary', *Inquiry* 16, 1973, 95–100.

73 See in particular N. Evernden, *The Natural Alien*, Toronto and Buffalo: University of Toronto Press, 1985. Also, E. M. Curley, 'Man and nature in Spinoza', in J. Wetlesen (ed.) *Spinoza's Philosophy of Man*, Proceedings of the Scandinavian Spinoza Symposium 1977, Oslo: Universiktsforlaget, 1978, pp. 19–26.

74 In e.g. A. Brennan, *Thinking about Nature*, Athens, GA: University of Georgia Press, 1988, especially the chapter on 'Theory, fact and value'. Deep Ecology does not go far enough for G. Foley, 'Deep ecology and subjectivity', *The Ecologist* 18 (4/5), 1988, 119–22.

75 The chief exegete in the environmental field is M. E. Zimmerman. See e.g. 'Towards a Heideggerian *ethos* for radical environmentalism', *Environmental Ethics* 5, 1983, 99–131; 'The critique of natural rights and the search for a non-anthropocentric basis for moral behaviour', *J. Value Inquiry* 19, 1985, 19–43.

76 There are echoes here of the Gaia hypothesis. I sometimes feel that Heidegger was putting poetically what science has subsequently revealed but I expect to be told that I have an imperfect understanding of the depth of Heidegger's thought.

77 Quoted by L. Westra, 'Let it be: Heidegger and future generations', *Environmental*

Ethics 7, 1985, 341–50, from Heidegger's *Basic Writings*, New York: Harper and Row, 1977, ed. D. F. Krell.

78 The staunchest protagonist of the continued primacy of western rationalism is generally held to be John Passmore, aided by the clarity of his expression. However, in the second edition of *Man's Responsibility for Nature* (London: Duckworth, 1980) he says in an Appendix that the working out of a new metaphysics for nature is 'the most important task that lies ahead of philosophy' (p. 215).

79 J. B. Callicott, 'Intrinsic value, quantum theory, and environmental ethics', *Environmental Ethics* 7, 1985, 257–75. This identification presumably goes beyond the dualities derived by some authors from the 'humans live in two worlds' (ecological and psychological) stance where it is argued that 'nature' cannot be an ethical realm in the way 'human society' can: there are more differences than similarities between the two realms in this view. See for example R. W. Gardiner, 'Between two worlds: humans in nature and culture', *Environmental Ethics* 12, 1990, 339–52.

80 The notion of self-realisation extending beyond the human is fundamental to F. Matthews, *The Ecological Self*, London: Routledge, 1991. She develops in particular the version of it called the **conatus** by Baruch Spinoza (1632–77).

81 Though we should not forget that F. Capra's *The Tao of Physics*, Berkeley, CA: Shambhala Press/London: Wildwood House, 1975, was a best-seller and there have been several imitators, such as G. Zukav, *The Dancing Wu Li Masters*, London: Fontana, 1979; R. Jones, *Physics as Metaphor*, London: Abacus Books, 1983.

82 C. Spretnak, *The Spiritual Dimension of Green Politics*, Santa Fé, NM: Bear and Co., 1986.

83 P. S. Wenz, *Environmental Justice*, Albany, NY: SUNY Press, 1988; L. K. Caldwell, *International Environmental Policy: Emergence and Dimensions*, Durham, NC: Duke University Press, 1991, 2nd edn.

84 The nexus between philosophy, economics and law is made (in a largely North American context) by M. Sagoff, *The Economy of the Earth*, Cambridge University Press Studies in Philosophy and Public Policy, 1988. Examples of the introduction of environmental considerations into that most active of conflicts, warfare, can be seen in e.g. G. Plant (ed.), *Environmental Protection and the Law of War*, London: Pinter, 1990; J. Käkönen (ed.), *Perspectives on Environmental Conflict and International Politics*, London: Pinter (Tampere Peace Research Institute Series).

85 See for example the discussion in L. D. Guruswamy, 'Laws controlling mercury pollution: issues and implications for the United Kingdom', *Lloyds Maritime and Commercial Law Quarterly* (Nov. 1982 part 4), 1–25.

86 L. D. Guruswamy, 'Environmental management in a North Sea coastal zone: law, institutions and policy', *Natural Resources Journal* 25, 1985, 233–42. This is put in a wider context by the same author in 'Waste management planning', *J. Env. Management* 21, 1985, 69–84.

87 L. D. Guruswamy, *op. cit.*, 1982.

88 P. C. Yeager, *The Limits of Law: the Public Regulation of Private Pollution*, Cambridge: Cambridge University Press, 1991.

89 Commission of the European Communities, Draft Resolution of the Council, *Fourth Environmental Action Programme of the EEC (1987–92)*, Luxembourg: COM (86) 485 final, 9:10:86, document CB-CO-86-476-EN-C.

90 D. M. Johnston, 'Marine pollution agreements: successes and problems', in J. E. Carroll (ed.) *International Environmental Diplomacy: The Management and*

Resolution of Transfrontier Environmental Problems, Cambridge: Cambridge University Press, 1988, pp. 199–206. For a wider treatment see L. K. Caldwell, *International Environmental Policy: Emergence and Dimensions*, Durham, NC: Duke University Press, 1984.

91 J. E. Carroll, 'Conclusion', in J. E. Carroll (ed.), *op. cit.*, 1988, pp. 275–79.
92 R. J. Johnston, 'Laws, states and super-states: international law and the environment', *Applied Geography* 12, 1992, 211–28.
93 For examples of books dealing with international law in this context, see the two volumes edited by René-Jean Dupuy, *The Settlement of Disputes on the New Natural Resources*, The Hague: Nijhoff, 1983, and *The Future of the International Law of the Environment*, The Hague: Nijhoff, 1985. See also chapter 8 ('Shaping world institutions') of L. K. Caldwell, *Between Two Worlds. Science, the Environmental Movement, and Policy Choice*, Cambridge: Cambridge University Press, 1990.
94 Or even beyond, as in Henryk Skolimowski's attempt to formulate a new worldview, 'From cosmology to consciousness', in *Living Philosophy. Eco-philosophy as a Tree of Life*, London: Arkana Books, 1992.

6 ONLY THE ROAD . . .

1 'Travellers! There is no road but the travelling.'
2 K. Davies, 'What is ecofeminism?', *Women and Environments* 10 (3), 1988, 4–6. A more accessible introduction and summary is C. Merchant, 'Earthcare', *Environment* 23 (5), 1981, 6–13, 38, 40. A broader insight is gained from I. Diamond and G. F. Orenstein, *Reweaving the World: the Emergence of Ecofeminism*, San Francisco: Sierra Club Books, 1990.
3 'The conceptual connections between the dual dominations of women and nature are located in an oppressive patriarchal conceptual framework characterized by a logic of domination', K. J. Warren, 'The power and the promise of ecological feminism', *Environmental Ethics* 12, 1990, 125–46.
4 Some of the most useful writings include, besides those given in other footnotes to this section, D. LaChapelle, *Earth Wisdom*, Silverton, CO: Finn Hill Arts, 1978; R. R. Reuther, *New Woman, New Earth*, New York: Seabury Press, 1975; S. Griffin, *Women and Nature*. New York: Harper and Row, 1978; L. Caldecott and S. Leland (eds), *Reclaim the Earth*, London: Women's Press, 1983. There is a special number of *Environmental Review* (8 (1), 1984) entitled 'Women and environmental history', with a good bibliography by J. Yett, 86–94.
5 C. Merchant, *The Death of Nature: Women, Ecology and the Scientific Revolution*, San Francisco: Harper and Row, 1980. See also the compact historical discussion (pp. 272–7) in L. Schiebinger, *The Mind has no Sex? Women in the Origin of Modern Science*, Cambridge, MA and London: Harvard University Press, 1989.
6 For example, the work of Brian Easlea, *Science and Sexual Oppression: Patriarchy's Confrontation with Woman and Nature*, London: Weidenfeld and Nicholson, 1981; *idem, Fathering the Unthinkable: Masculinity, Scientists and the Arms Race*, London: Pluto Press, 1983.
7 K. Davies, *op. cit.*, p. 6.
8 V. Plumwood, 'Feminism and ecofeminism: beyond the dualistic assumptions of women, men and nature', *The Ecologist* 22, 1992, 8–13. This issue of *The Ecologist* (22 (1) is devoted to feminism, nature and development: pp. 33–5 are devoted to short accounts of books in the field. A recent example of sociopolitical feminism is M. Mellor, *Breaking the Boundaries: Towards a Feminism Green Socialism*, London: Virago Press, 1992.

9 K. Warren, 'Feminism and ecology: making connections', *Environmental Ethics* **9**, 1987, 3–20.

10 S. Prentice, 'Taking sides: what's wrong with ecofeminism?', *Women and Environments* **10** (3), 1988, 9–10.

11 *op. cit.*, 1981. A discussion of the possibility of a feminist science can be found in chapter 9 of H. Longino, *Science as Social Knowledge*, Princeton, NJ: Princeton University Press, 1990.

12 W. I. Thompson, *The Time Falling Bodies Take to Light: Mythology, Sexuality and the Origins of Culture*, London: Rider/Hutchinson, 1981.

13 See, though, V. L. Norwood, 'Heroines of nature: four women respond to the American landscape', *Environmental Review* **8**, 1984, 34–56.

14 In scientific terms, these are very difficult words to handle. 'Balance' implies some condition of long-term stability which is rarely present even in natural ecosystems; 'stability' can be defined in several different ways and is not consistently used in the literature.

15 This is not an easy term, either. See the discussion in e.g. M. Redclift, *Sustainable Development. Exploring the Contradictions*, London and New York: Methuen, 1987. We might argue that non-renewable resources should be expended only as capital investment into future systems which are renewable ones; in the case of fossil fuels for instance into learning how to do without them.

16 This is very difficult because of the legal convention that interests and rights can only be accorded to individuals and not collections such as species, biological communities and ecosystems. Thus an individual elephant might have rights but the species elephant cannot. See M. Sagoff, *The Economy of the Earth: Philosophy, Law and the Environment*, Cambridge: Cambridge University Press, 1988, ch. 7.

17 This is put into the context of a search for a non-duality arising out of but not confined to modern science by M. E. Zimmerman, 'Quantum theory, intrinsic value, and panentheism', *Environmental Ethics* **10**, 1988, 3–30.

18 Additional references include Arne Naess, 'Deep Ecology and ultimate premises', *The Ecologist* **18**, 1988, 128–31. This is part of a special issue of *The Ecologist* on the subject of Deep Ecology, both pro and con. The last paper, E. Goldsmith, 'The way: an ecological world-view', 160–85, is a set of 67 rules for the ecosphere and uses 'Gaian principles' a great deal.

19 For example, R. Sylvan, 'A critique of Deep Ecology', *Radical Philosophy* **40**, 2–12 and **41**, 10–22, and several contributions in *The Ecologist* issue in n 16, *supra*.

20 Consider also the problems of how we come to know about the planet's ecology: beyond mysticism, there are only sentences and images which communicate information to us. Thus the state of the world 'resonates' in society via various mechanisms and creates anxieties which are then, sometimes, addressed rationally. See N. Luhmann, *Ecological Communication*, Cambridge: Polity Press, 1989.

21 A. Berque, *Le Sauvage et L'artifice: les Japonais devant la Nature*, Paris: Gallimard, 1985; *idem*, 'Some traits of Japanese *fūdosei*', *The Japan Foundation Newsletter* **14** (5) 1987, 1–7.

22 M. Watanabe, 'The conception of nature in Japanese culture', *Science* **183**, 1974, 279–82. See also the more complex discussions of Asian thought in J. B. Callicott and R. T. Ames (eds), *Nature in Asian Traditions of Thought: Essays in Environmental Philosophy*, Albany, NY: SUNY Press, 1989, and especially H. Tellenbach and B. Kimura, 'The Japanese concept of "nature"', pp. 163–82; D. E. Shaner, 'The Japanese experience of nature', pp. 183–209.

23 Hwa Yol Yung, 'The ecologic crisis: a philosophic perspective, East and West', *Bucknell Review* 20, 1972, 25–44. He draws our attention to such works as P. P. Weiner (ed.), *Ways of Thinking of Eastern Peoples*, Honolulu: East-West Centre Press, 1964, and the works of D. T. Suzuki, especially *Zen and Japanese Culture*, New York: Pantheon Books, 1959.

24 There are echoes here of Thoreau, who wrote of the earth as 'living poetry' and of Heidegger, for whom poetry was the essence of truth. A comparison of Zen and Heidegger can be found in P. Kreeft, 'Zen in Heidegger's *Gelassenheit*', *International Philosophical Quarterly* 11, 1971, 521–45.

25 I. G. Simmons, 'The whiskey ceremony: humans and environments in Japan and Great Britain', in M. Kurosaka (ed.) *Resonance in Nature*, Tokyo: Shisakusha Publishing Co., 1990, vol. 3: *Towards a Creative Relationship with the Earth*, pp. 131–59 (in Japanese).

26 B. L. Whorf (ed. J. B. Carroll), *Language, Thought and Reality*, London: Chapman and Hall, 1956.

27 Set out with clarity in a theological context but with some environmental allusions ('Green fundamentalism is thus a delusion', p. 151), in D. Cupitt, *Creation out of Nothing*, London: SCM Press, 1990. For a hard-felt counter-view see G. Steiner, *Real Presences: Is There Anything in What We Say?*, London and Boston: Faber and Faber, 1989, p. 227, 'The limits of our language are not . . . those of our world', though at that moment he is discussing poetry, art and music. But if one form of *poesis*, why not others?

28 A metaphor, of course. But take more than one mirror and we can have a labyrinth where any image (including that of the self) dissolves into parody. So mirrors can be a metaphor in the discourses of postmodernism. See chapter 7 of R. Kearney, *The Wake of Imagination: Toward a Postmodern Culture*, Minneapolis: University of Minnesota Press, 1988.

29 The intellectual precursor of Derrida in these considerations of metaphor, undecidability and self-reflexivity was Nietzsche (1844–1900): 'What, therefore, is truth? A mobile army of metaphors, metonymies, anthropomorphisms; truths are illusions of which one has forgotten they are illusions.' (Note the metaphor of 'mobile army'.)

30 Comprehension of these ideas is not easy. I am especially indebted to J. Hoy, 'Jacques Derrida', in Q. Skinner, *op. cit.*, 1985, pp. 41–64; ch. 2 of Maran Sarup, *op. cit.*, 1988; C. Norris, *Derrida*, London: Fontana, 1987. For a specific extension of these types of ideas to the environment, see J. Cheney, 'Postmodern environmental ethics: ethics as bioregional narrative, *Environmental Ethics*' 11, 1989, 117–34.

31 A search has also been made for the 'middle voice' of Sanskrit and Greek, which is neither active nor passive. Such a voice, were it present in modern languages, might help us to avoid subjectivity as pure agency or pure passivity and also to avoid treating 'the environment' in a merely instrumental way. See J. Llewelyn, *The Middle Voice of Ecological Conscience: A Chiasmic Reading of Ecological Responsibility in the Neighbourhood of Levinas, Heidegger and Others*, London: Macmillan, 1991.

32 The metaphor is from P. Scott, 'The postmodern challenge. III. Reaching beyond enlightenment', *The Times Higher Education Supplement*, 24 August 1980, 28. His is the succeeding quotation also.

33 J. Cheney, *op. cit.*, 1989, see n 28.

34 Truth being taken as an accurate representation of some aspect of the cosmos.

35 For example, Q. Skinner deals in his edited book *The Return of Grand Theory in the Human Sciences* (Cambridge University Press, 1985) with eight individuals

and one School. It is only the latter (the *Annales* historians) which treat with it at all.

36 R. Bhaskar, *The Possibility of Naturalism: A Philosophical Critique of the Contemporary Human Sciences*, Brighton, Sussex: Harvester Press, 1979.

37 Structuralism is particularly associated with the French anthrolopogist Claude Lévi-Strauss (1908–), and structuration with the British sociologist Anthony Giddens (1938–). Works of especial utility are M. Glucksmann, *Structural Analysis in Contemporary Social Thought*, London: RKP, 1974; A. Giddens, *The Constitution of Society: Outline of the Theory of Structuration*, Cambridge, Polity Press, 1984.

38 N. Smith, *Uneven Development: Nature, Capital and the Production of Space*, Oxford: Blackwell, 1990, 2nd edn.

39 Smith is quoted approvingly, among others, by M. Fitzsimmons as arguing that primordial and separate nature cannot exist. 'Nature' is only ontologically possible as part of a relationship with human society. M. Fitzsimmons, 'The matter of nature', *Antipode* 21, 1989, 106–20.

40 P. B. Medawar, *Pluto's Republic*, Oxford: Oxford University Press, 1982. More extreme statements are made: e.g., N. Cartwright, *How the Laws of Physics Lie*, Oxford: Clarendon Press, 1983: ' There is only what happens and what we say about it. Nature tends to a wild profusion, which our thinking does not wholly confine. We do not know whether we are in a tidy universe or an untidy one. I imagine that natural objects are much like people in societies. Their behaviour is constrained by some specific laws and by a handful of general principles. But it is not determined in detail, even statistically.'

41 I. Hacking, *Representing and Intervening: Introductory Topics in the Philosophy of Natural Science*, Cambridge: Cambridge University Press, 1983. Compare with Marx's dictum that the point is not to understand the world but to change it.

42 P. B. Medawar, *The Limits to Science*, New York: Harper and Row, 1984.

43 In which we recognise the word 'poetry'.

44 P. B. Medawar, *Pluto's Republic*, Oxford: Oxford University Press, 1982.

45 P. B. Medawar, *The Art of the Soluble: Creativity and Originality in Science*, Harmondsworth: Penguin, 1967; J.-F. Lyotard, *The Postmodern Condition: A Report on Knowledge*, Minneapolis: University of Minnesota Press, Theory and History of Literature vol. 10, 1984 (originally published in French in 1974).

46 Ideally, that is. Scientific fraud in highly competitive areas like biomedicine has been detected with increasing frequency in recent years, and the activities of some of the drug companies are scarcely beyond reproach in scientific terms, let alone ethical ones.

47 S. Yearley, 'Greens and science: a doomed affair?', *New Scientist* 131 (1777, 13 July 1991), 37–40.

48 See chapter 10 ('On the ethical neutrality of science – the indictment of the scientific mentality') of B. Easlea, *Liberation and the Aims of Science* (London: Chatto and Windus for the University of Sussex Press, 1973) for a polemic on science as an instrument of control and capitalism.

49 What follows is dependent in large measure on but also adapted from K. E. Peters, 'Humanity in nature: conserving yet creating', *Zygon* 24, 1989, 469–85.

50 See A. Schnaiberg, *The Environment from Surplus to Scarcity*, New York: Oxford University Press, 1980. There is of course a personal level in this as well: consider the way in which Goethe's Dr Faustus is archetypally divided between the rational intellectual and the emotional and sensual. Mind and matter are for ever irreconcilable for him.

51 There is an interesting account of modern fallacies in C. Birch, 'Eight fallacies

of the modern world and five axioms for a postmodern worldview', *Perspectives in Biology and Medicine* **32**, 1988, 12–30.

52 R. Ornstein and P. Ehrlich, *New World New Mind. Changing the Way we Think to Save our Future*, London: Methuen, 1989.

53 An unpleasant analogy. A live frog dropped into a pan of boiling water might just manage to react so swiftly as to jump out; a frog put into cold water which is then heated to boiling never perceives that the moment has come to leap out.

54 A. Wilder, *The Rules are no Game*, London and New York: Routledge, 1987.

55 From the collection *In Search of the Far Side*, London: Futura Publications (not paginated: would be p. 40).

56 R. Ornstein and P. R. Ehrlich, *Conscious Evolution, op. cit.*, p. 73.

57 Balachandra Rajan, *The Overwhelming Question: A Study of the Poetry of T. S. Eliot*, Toronto and Buffalo: University of Toronto Press, 1976. The 'question' was the extent to which the parts could only be understood if the whole was also comprehensible.

58 Paul Shepherd argues that the home of western thought was the desert edge in the time of the empires of Egypt and Assyria, habitat of 'the peoples of the dry landscapes' (*Nature and Madness*, San Francisco: Sierra Club Books, 1982, ch. 3).

59 There is a discussion of this in a review by J. Gray of E. and A. Margalit (eds), *Isaiah Berlin: A Celebration* (London: Hogarth Press, 1990) in the *Times Literary Supplement*, 5 July 1991, 3.

60 The Greek Archilocus (*f.l. c.* 650 BC): 'The fox knows many things but the hedgehog knows one big thing.' In his essay on this saying, Isaiah Berlin (*The Hedgehog and the Fox: An Essay on Tolstoy's View of History*, London: Weidenfeld and Nicholson, 1953) suggests it marks one of the deepest differences which divides writers and thinkers and possibly humans in general. Hedgehogs include Dante, Plato, Nietzsche and Proust; foxes are exemplified by Aristotle, Shakespeare and Joyce.

61 I have in mind for the East the notion of the *Tao* as expressed in the symbol in which black and white occupy equal interpenetrative portions of a circle but each contains a small mini-circle of the other. For the West, the words of the fourteenth-century mystic Julian of Norwich: 'Sin is behovely, but all shall be well . . . and all manner of thing shall be well.'

62 This is hinted at by S. Toulmin, 'The fire and the rose', in *The Return to Cosmology*, Berkeley and Los Angeles: University of California Press, 1982, pp. 255–74 and made more explicit (largely following Heidegger) by J. Grange, 'On the way towards foundational ecology', *Soundings* **60**, 1977, 135–49.

63 S. J. Gould, *Wonderful Life: The Burgess Shale and the Nature of History*, London: Hutchinson, 1989, pp. 284–5.

64 The process philosophy of A. N. Whitehead (1861–1947) may be an under-explored resource here.

65 I. Burton and R. W. Kates, 'The Great Climacteric, 1798–2048', in R.W. Kates and I. Burton (eds) *Geography, Resources, Environment*, Vol. II: *Themes from the Work of Gilbert F. White*, Chicago: University of Chicago Press, 1986, pp. 339–60. Also the further discussion by R. W. Kates, 'Theories of nature, society and technology', in E. Baark and U. Svedin (eds) *Man, Nature and Technology: Essays on the Role of Ideological Perceptions*, London: Macmillan, 1988, pp. 7–36.

66 Which they get in D. E. Cooper, 'The idea of environment', in D. E. Cooper and J. A. Palmer (eds) *The Environment in Question: Ethics and Global Issues*, London and New York: Routledge, 1992, pp. 165–80.

67 I am tempted to say that being laid-back is more important but no doubt would
be accused of frivolity. 'Laid-open' conveys something of the sense I want to
impart. The formulation of the abandonment of Utopian ideals comes from Isaiah
Berlin, *The Crooked Timber of Humanity: Chapters in the History of Ideas*,
London: Fontana Press, 1991.

FURTHER READING

1 INTRODUCTION

This chapter lays out the ground for further discussion, so many of the works referred to in the Notes recur later. This selection is thus highly restricted.

Empirical history

I. G. Simmons, *Changing the Face of the Earth*, Oxford: Blackwell, 1989.
B. L. Turner *et al.* (eds), *The Earth as Transformed by Human Action*, Cambridge University Press/Clark University, 1990.
C. Ponting, *A Green History of the World*, London: Sinclair-Stevenson, 1991.
M. Oelschlager, *The Idea of Wilderness from Prehistory to the Age of Ecology*, New Haven and London: Yale University Press, 1991.

Culture and society

N. Luhmann, *Ecological Communication*, Cambridge: Polity Press, 1989, trans. J. Bednarz. (First published in German in 1986.)
K. Lee, *Social Philosophy and Ecological Scarcity*, London: Routledge, 1989.
P. L. Berger and T. Luckmann, *The Social Construction of Reality, A Treatise in the Sociology of Knowledge*, London: Allen Lane, 1967.
L. Winner, *Autonomous Technology*, Cambridge, MA: MIT Press, 1967.
L. Winner, *The Whale and the Reactor*, University of Chicago Press, 1986.
J. Ellul, *The Technological Society*, New York: Knopf, 1964.
Yi-Fu Tuan, 'Our treatment of the environment in ideal and actuality', *American Scientist* 58, 1970, 244–9.
J. R. Short, *Imagined Country: Society, Culture and Environment*, London and New York: Routledge, 1991.
C. Merchant, *Ecological Revolutions, Nature, Gender, and Science in New England*, London and Chapel Hall, NC: University of North Carolina Press, 1989.
S. Cotgrove, *Catastrophe or Cornucopia: the Environment, Politics and the Future*, Chichester: Wiley, 1982.
T. O'Riordan, 'The challenge for environmentalism', in R. Peet and N. Thrift (eds), *New Models in Geography*, London: Unwin Hyman, 1989, vol. II, 77–102.
N. Everden, *The Social Creation of Nature*, Baltimore: Johns Hopkins Press, 1992.
C. Merchant, *Radical Ecology. The Search for a Liveable World*, New York: Routledge, 1992.

FURTHER READING

Ethics and values

R. Attfield, *The Ethics of Environmental Concern*, Oxford: Blackwell, 1983.
T. Attig and D. Scherer (eds), *Ethics and the Environment*, Englewood Cliffs, NJ: Prentice Hall, 1983.

Non-western ideas

J. D. Hughes, *American Indian Ecology*, El Paso, TX: Texas Western Press, 1983.
J. B. Callicott and R. T. Ames (eds), *Nature in Asian Traditions of Thought*, Albany, NY: SUNY Press, 1989.
M. E. Tucker, 'The relevance of Chinese Neo-Confucianism for the reverence of nature', *Environmental History Review* **15**, 1991, 55–69.
A. Berque, *Le Sauvage et l'Artifice: Les Japonais devant la Nature*, Paris: Gallimard, 1986.

2 THE NATURAL SCIENCES AND TECHNOLOGY

General

J. Ziman, *Reliable Knowledge: An Exploration of the Grounds for Belief in Science*, Cambridge: Cambridge University Press, 1978.
A. F. Chalmers, *What is this Thing Called Science?*, Milton Keynes: Open University Press, 1982.
A. F. Chalmers, *Science and its Fabrication*, Milton Keynes: Open University Press, 1990.
H. Longino, *Science as Social Knowledge*, Princeton, NJ: Princeton University Press.
D. Faust, *The Limits of Scientific Reasoning*, Minneapolis: University of Minnesota Press, 1990.
P. B. Medawar, *The Art of the Soluble: Creativity and Originality in Science*, Harmondsworth: Penguin Books, 1969.
M. Midgley, *Science as Salvation: A Modern Myth and its Meaning*, London and New York: Routledge, 1992.

Biology and ecology

G. E. Hutchinson, *The Ecological Theatre and the Evolutionary Play*, New Haven, CT: Yale University Press.
R. P. McIntosh, *The Background of Ecology: Concept and Theory*, Cambridge: Cambridge University Press, 1985.
S. J. Gould, *Wonderful Life: The Burgess Shale and the Nature of History*, London: Hutchinson, 1989.
J. E. Lovelock, *The Ages of Gaia: A Biography of our Living Earth*, Oxford: Oxford University Press, 1988.
T. Ingold, *Evolution and Social Life*, Cambridge: Cambridge University Press, 1986.

Other views

J. Douglas and A. Wildavsky, *Risk and Culture: An Essay on the Selection of Technological and Environmental Dangers*, Berkeley and Los Angeles: University of California Press, 1982.
I. Stewart, *Does God Play Dice? The Mathematics of Chaos*, Oxford: Blackwell, 1989.

E. Jantsch, *The Self-organizing Universe: Scientific and Human Implications of the Emerging Paradigm of Evolution*, Oxford: Pergamon Press, 1980.

I. Prigogine and I. Stengers, *Order out of Chaos: Man's New Dialogue with Nature*, London: Flamingo Books, 1985.

A. Henderson-Sellers, 'Climate, models and geography', in B. Macmillan (ed.) *Remodelling Geography*, Oxford: Blackwell, pp. 117–46.

3 SOCIETY AND ENVIRONMENT: MATERIALIST INTERPRETATIONS

Specific social sciences

D. W. Pearce, A. Markandya and E. B. Barbier, *Blueprint for a Green Economy*, London: Earthscan Publications, 1989.

W. U. Chandler, *The Changing Role of the Market in National Economies*, Washington, DC: Worldwatch Paper 72.

P. Earl, *The Economic Imagination. Towards a Behavioural Analysis of Choice*, Brighton: Wheatsheaf Press, 1983.

J. R. L. Proops, 'Ecological economics: rationale and problem areas', *Ecological Economics* 1, 1989, 59–76.

A-M. Jansson (ed.), *Integration of Economy and Ecology: An Outlook for the Eighties*, Stockholm: University of Stockholm Askö Laboratory, 1984.

V. Anderson, *Alternative Economic Indicators*, London and New York: Routledge, 1991.

J. Martinez-Alier, *Ecological Economics. Energy, Environment and Society*, Oxford: Blackwell, 1987.

E. B. Barbier, *Economics, Natural Resource Scarcity and Development: Conventional and Alternative Views*, London: Earthscan, 1989.

V. Anderson, *Alternative Economic Indicators*, London: 1991.

A. Dobson, *Green Political Thought: An Introduction*, London: Routledge, 1990.

A. Atkinson, *Principles of Political Ecology*, London: Belhaven Press, 1991.

S. Cotgrove, *Catastrophe or Cornucopia? The Environment, Politics and the Future*, Chichester: Wiley, 1982.

T. Roszak, *Person/Planet: The Creative Disintegration of Industrial Society*, Garden City, NY: Anchor Press, 1978/London: Granada Books, 1981.

J. Porritt, *Seeing Green: The Politics of Ecology Explained*, Oxford: Blackwell, 1984.

C. Spretnak and F. Capra, *Green Politics: The Global Promise*, New York: Dutton, 1984/London: Paladin Books, 1986.

D. Wall, *Getting There: Steps to a Green Society*, London: Green Print, 1990.

R. W. Kates, 'The human environment: the road not taken, the road still beckoning', *Ann. Assoc. Amer. Geogr.* 77, 1987, 525–34.

P. Burnham and R. F. Ellen (eds), *Social and Ecological Systems*, London: Academic Press, 1979.

R. F. Ellen, *Environment, Subsistence and System*, Cambridge: Cambridge University Press, 1982.

J. W. Bennett, *The Ecological Transition: Cultural Anthrolopology and Human Adaptation*, Oxford: Pergamon Press, 1976.

D. L. Hardesty, *Ecological Anthropology*, New York: Wiley, 1977.

L. Allison, *Environmental Planning: A Political and Philosophical Analysis*, London: Allen and Unwin, 1974.

Landscape evaluation; environmental hazards

J. Appleton, *The Experience of Landscape*, London: Wiley, 1975.

J. Appleton, *The Symbolism of Habitat: An Interpretation of Landscape in the Arts*, Seattle, WA: University of Washington Press, 1990.

R. W. Kates and I. Burton (eds), *Geography, Resources and Environment*, Vol. II, *Themes from the Work of Gilbert F. White*, Chicago and London: University of Chicago Press, 1986, especially the essays by H. Kunreuther and P. Slovic, 'Decision-making in hazard and resource management', pp. 153–87; A. V. T. Whyte, 'From hazard perception to human ecology', pp. 240–71; T. O'Riordan, 'Coping with environmental hazards', pp. 272–309.

K. Hewitt, 'The idea of calamity in a technocratic age', in K. Hewitt (ed.) *Interpretations of Calamity*, London: Allen and Unwin, 1983, Risks and Hazards series No. 1, 3–32.

E. Waddell, 'The hazards of scientism: a review article', *Human Ecology* 5, 1977, 69–76.

Social sciences – general

A. Schnaiberg, *The Environment from Surplus to Scarcity*, New York and Oxford: Oxford University Press, 1980.

A. Schmidt, *The Concept of Nature in Marx*, London: New Left Review, 1971, trans. B. Fowkes (German original publ. 1962).

D. Owen, *Green Reporting: Accountancy and the Challenge of the Nineties*, London: Chapman and Hall, 1991.

L. Winner, *The Whale and the Reactor. A Search for Limits in an Age of High Technology*, Chicago and London: University of Chicago Press, 1986.

S. Yearsley, *The Green Case. A Sociology of Environmental Issues, Arguments and Politics*, London: Harper Collins Academic, 1991.

W. Leiss, *The Limits to Satisfaction: An Essay on the Problem of Needs and Commodities*, Toronto and Buffalo: University of Toronto Press, 1976.

4 THE LIFEWORLD

Perception, cognition and the lifeworld

Yi Fu Tuan, *Topophilia: A Study of Environmental Perception, Attitudes, and Values*, Englewood Cliff, NJ: Prentice-Hall, 1974.

H. Spiegelberg, *Doing Phenomenology*, The Hague: Nijhoff, 1975.

A. Schutz, *The Phenomenology of the Social World*, London: Heinemann, 1972 (first publ. in German in 1932).

C. Rose, 'Wilhelm Dilthey's philosophy of historical understanding. A neglected heritage of contemporary humanistic geography', in D. R. Stoddart (ed.) *Geography, Ideology and Social Concern*, Oxford: Blackwell, 1981, pp. 99–133.

N. Everden, *The Natural Alien. Humankind and Environment*, Toronto, Buffalo and London: University of Toronto Press, 1985.

D. Abram, 'Merleau-Ponty and the voice of the Earth', *Env. Ethics* 10, 1988, 101–20.

W. J. Mills, 'Positivism reversed: the relevance of Giambattista Vico', *Trans. Inst. Brit. Geogr.* (NS) 7, 1982, 1–14.

L. Pompa, *Vico: A Study of the New Science*, Cambridge: Cambridge University Press, 1975.

The fine arts

K. Clark, *Landscape into Art*, London: John Murray, 1949, and later editions.

P. Shepard, *Man in the Landscape: A Historic View of the Esthetics of Nature*, New York: Knopf, 1976.

E. Carli, *The Landscape in Art*, New York: William Morrow, 1980 (first published in Italy in 1979).

A. R. Turner, *The Vision of Landscape in Renaissance Italy*, Princeton University Press, 1966 (paperback edn 1974).

R. Cafritz, L. Gowing and D. Rosand, *Places of Delight: The Pastoral Landscape*, Washington, DC: The Phillips Collection, 1988.

F. Goodwin and T. Hughes, *Remains of Elmet*, London: Faber and Faber, 1979.

A. Kolodny, *The Lay of the Land: Metaphor as Experience and History in American Life and Letters*, Chapel Hill, NC: University of North Carolina Press, 1975.

D. C. D. Pocock (ed.), *Humanistic Geography and Literature: Essays on the Experience of Place*, London: Croom Helm, 1981.

R. Paulson, *Literary Landscapes: Turner and Constable*, New Haven: Yale University Press, 1982.

J. Bate, *Romantic Ecology: Wordsworth and the Environmental Tradition*, London and New York: Routledge, 1991.

R. Murray Schaefer, *The Tuning of the World*, New York: Knopf, 1977.

D. Trussell, 'The arts and planetary survival' (Part I), *The Ecologist* 19, 1989, 179–5; (Part II), *The Ecologist* 20, 1990, 4–8.

J. F. Wohlwill, 'The place of order and uncertainty in art and environmental aesthetics', *Motivation and Emotion* 4, 1980, 133–42.

Critical theory

D. Harvey, *The Condition of Postmodernity*, Oxford: Blackwell, 1990.

Z. Bauman, *Intimations of Postmodernity*, London and New York: Routledge, 1992.

5 NORMATIVE BEHAVIOUR

General

E. Katz, 'Ethics and philosophy of the environment: a brief review of the major literature', *Environmental History Review* 15, 1991, 79–86.

K. Thomas, *Man and the Natural World: Changing Attitudes in England 1500–1800*, London: Allen Lane, 1983.

D. C. Pitt (ed.), *The Future of the Environment. The Social Dimensions of Conservation and Ecological Alternatives*, London and New York: Routledge, 1988.

Philosophy and ethics

E. C. Hargrove, *Foundations of Environmental Ethics*, Englewood Cliffs, NJ: Prentice Hall, 1989.

K. S. Shrader-Frechette (ed.), *Environmental Ethics*, Pacific Grove, CA: The Boxwood Press, 1981.

F. Matthews, *The Ecological Self*, London: Routledge, 1991.

R. Elliot and A. Gare (eds), *Environmental Philosophy*, University Park and London: Pennsylvania State University Press, 1983.

R. Attfield, *The Ethics of Environmental Concern*, Oxford: Blackwell, 1983.

J. Passmore, *Man's Responsibility for Nature*, London: Duckworth, 1980, 2nd edn.

J. B. Callicott and R. T. Ames (eds), *Nature in Asian Traditions of Thought*, Albany, NY: SUNY Press, 1989.

A. Brennan, *Thinking about Nature*, Athens, GA: University of Georgia Press, 1988.

J. B. Callicott, 'Intrinsic value, quantum theory, and environmental ethics', *Environmental Ethics* 7, 1985, 257–75.

J. R. Engel and J. G. Engel (eds), *Ethics of Environment and Development*, London: Belhaven Press, 1990.

Religion

J. D. Hughes and J. Swan, 'How much of the earth is sacred space?', *Environmental Review* 10, 1986, 247–59.

J. Kay, 'Human domination of nature in the Hebrew Bible', *Ann. Assoc. Amer. Geogr.* 79, 1989, 214–32.

J. Cobb, *Ecology and Religion: Toward a New Christian Theology of Nature*, Ramsey, NJ: Paulist Press, 1983.

M. Fox, *Creation Sprituality*, London: Harper Collins, 1991.

A. Primavesi, *From Apocalypse to Genesis*, London: Burns and Oates, 1991.

C. C. Park, *Caring for Creation*, London: Marshall Pickering, 1992.

Law and policy

P. S. Wenz, *Environmental Justice*, Albany, NY: SUNY Press, 1988.

L. K. Caldwell, *International Environmental Policy: Emergence and Dimensions*, Durham, NC: Duke University Press, 1991, 2nd edn.

M. Sagoff, *The Economy of the Earth*, Cambridge University Press, Studies in Philosophy and Public Policy, 1988.

C. Stone, *Should Trees Have Standing? Towards Legal Rights for Natural Objects*, Portola Valley, CA: Tioga Publishing Co., 1988, 2nd edn.

Wider context

H. Skolimowski, *Living Philosophy. Eco-philosophy as a Tree of Life*, London: Arkana Books, 1992.

A. Naess, *Ecology, Community and Lifestyle. Outline of an Ecosophy*, Cambridge: Cambridge University Press, 1989.

6 ONLY THE ROAD ...

General

R. Bhaskar, *The Possibility of Naturalism: A Philosophical Critique of the Contemporary Human Sciences*, Brighton, Sussex: Harvester Press, 1979.

I. Hacking, *Representing and Intervening: Introductory Topics in the Philosophy of Natural Science*, Oxford: Oxford University Press, 1983.

Feminism

I. Diamond and G. F. Orenstein, *Reweaving the World: The Emergence of Ecofeminism*, San Francisco: Sierra Club Books, 1990.

C. Merchant, *The Death of Nature: Women, Ecology and the Scientific Revolution*, San Francisco: Harper and Row, 1980.

B. Easlea, *Science and Sexual Oppression: Patriarchy's Confrontation with Women and Nature*, London: Weidenfeld and Nicholson, 1981.

Critical thinking

A. Naess, *Ecology, Community and Lifestyle*, Cambridge: Cambridge University Press, 1989.

P. Kreeft, 'Zen in Heidegger's *Gelassenheit*', *International Philosophical Quarterly* 11, 1971, 521–45.

J. Cheney, 'Postmodern environmental ethics: ethics as bioregional narrative', *Environmental Ethics* 1989, 117–34.

N. Smith, *Uneven Development: Nature, Capital and the Production of Space*, Oxford: Blackwell, 1990, 2nd edn.

M. Redclift, *Sustainable Development. Exploring the Contradictions*, London and New York: Methuen, 1987.

P. B. Medawar, *Pluto's Republic*, Oxford: Oxford University Press, 1982.

Broad-scale views

A. Schnaiberg, *The Environment from Surplus to Scarcity*, New York: Oxford University Press, 1980.

R. Ornstein and P. R. Ehrlich, *New World New Mind: Changing the Way we Think to Save our Future*, London: Methuen, 1989.

P. Sheperd, *Nature and Madness*, San Francisco: Sierra Club Books, 1982.

I. Berlin, *The Crooked Timber of Humanity: Chapters in the History of Ideas*, London: Fontana, 1990.

S. Toulmin, *The Return to Cosmology*, Berkeley and Los Angeles: University of California Press, 1982.

M. W. Lewis, *Green Delusions. An Environmentalist Critique of Radical Environmentalism*, Durham, NC/London: Duke University Press, 1992.

INDEX

Page numbers in italics refer to tables or illustrations.

Abram, David 81
acid rain 45
Adams, Ansel 92
aesthetics 70, 121
agency, and structure 154
ahimsa (non-injury) 132
anarchism 61
Anatomy of the World, An (Donne) 160
androcentrism 146
animal rights 124, 125
anthropic principle 2
anthropocentrism 123–4, 128, 147–8
anthropology 30, 65–7
Antipode, journal 64
Apocalypse Now (Coppola) 112
appropriate technology 131
architecture 115
arcology 115
Aristotle 126
arts, and environment: architecture 116; cinema 111–13; gardens 115–16; literary arts 93–109 (journalism 109–10, poetry 101–8, prose 94–101); music 109–11; television 113–14; visual arts 82–94 (painting and drawing 82–9, photography 91–3, sculpture 89–91)
Augustine, St 125
Avenue, Middleharnis (Hobbema) 85

Bacon, Francis 130
Bashō 107
Beagle journal (Darwin) 29, 30
Beauvoir, Simone de 144
behaviour, reflexive studies 77–81

behaviourism 7, 69–73
Being (Heidegger) 135–6
Bellini, Giovanni 84
Benedictine ethics 130
Bentham, Jeremy 68, 124, 182 (n90)
biocentric equality 134
biology 21
bioregion 57
biosphere 37, 125, 126
Bookchin, Murray 59, 61, 145
Brancusi, Constantin 90
Brave New World (Huxley) 97
Breugel, Pieter, the Elder 84
Buddhism 13, 132, 133, 148–9
Buddhist economics 52
Buson 107
butterfly effect 35

capitalism, and uneven development 154
carbon dioxide 27–8, 141
carrying capacity, of environment 66
Catastrophists 60, 79
centre–periphery model of development 63–4
Cézanne, Paul 89
chaos theory 35, 162
Chardin, Teilhard de 127, 128, 130
children's literature 96
China 13, 14, 18; *see also* Buddhism; Taoism
chlorofluorocarbons (CFCs) 141
Christianity 129–31
cinema, and the environment 111–13
Clare, John 101
class interests 57
climatology 27–8
cognition 76–7
communication 39–40, 81

science 18–21; and ecology 35–41; and
environment 40–1; positivism 19, 154;
as spectacle 155–8; and technology
4–6, 41–2; western 18; *see also* human
sciences; social sciences
science fiction 96–7
scientific method 171 (n4)
scientism 166 (n15)
scienza/coscienza 81
sculpture, and the environment 89–91
self, and the environment 78–81
self-organisation 59
self-realisation 134, 137, 195 (n80)
semiology 149–50
Seven Samurai, The (Kurosawa) 111
Sibelius, Jan 111
signifier, and signified 149, 150
Snow Country (Kawabata) 100–1
Snyder, Gary 105, 106–7, 131
social ecofeminism 146
social sciences 7–8; and environmentalist
thought 183 (n101); and natural
sciences 153; positivist 73–5
social theory of value 46, 60–1
socialist economics 48–50
sociology 7, 59–62
Soleri, Paolo 115
Sound of the Mountain, The (Kawabata)
102
space-ship earth 51
spacetime 121
Spinoza, Baruch 134
stewardship 56, 193 (n58)
stochastic behaviour 35, 174 (n63)
stock resources 45
stress ecology 25
structural organisation 36
structuralism 154, 199 (n37)
structuration 199 (n37)
structure and agency 154
subjective constructions 3–4
sustainability 48
Sutcliffe, Frank Meadows 91
Sutherland, Graham 89

Taoism 132, 200 (n61)
technical law 138–9
technology 20–1; appropriate 131; and
imperialism 42; and natural sciences
4–6; and social order 58; Third World
42
television, and the environment 113–14

telos 121–2, 158–9
Tempest, The (Giorgione) 84
Tess of the D'Urbervilles (Hardy) 97
theology, and ethical systems 128–31
Thomas, R. S. 102–3
thought-patterns, dominant 176 (n2)
Three Places in New England (Ives)
109–10
Thunen, A. von 63
time: in ecological terms 22; linearity 11,
127; perceptions 77–8; space and
biological complexity 25
Tolkien, J. R. R. 96
topophilia 77
traditions, non-literate 14
Train dans la Campagne (Monet) 87, 88
Train dans la Neige (Monet) 88
transboundary problems 118
transcendental signifier 150
transformative feminism 146
transnational law, EC 139–41
triage proposals 122
Turner, J. W. M. 85, 86

Ulysses (Joyce) 98–9
UN, and environmental policy 55
UN Law of the Sea 45, 140
universals, human 117
Universe symphony (Ives) 110
user values 47
Usines à Clichy (van Gogh) 88–9
utilitarianism 182 (n90)
Utopia 98, 162–3

values: consumer theory 46; ecological
37–40; economic 46–8; ethics 9, 168
(n32); and fact 162; inherent 120–1;
intrinsic 47; labour theory 49;
normative 118; options 47; user 47
Vico, Giambattista 81, 153
viriditas 130
visual arts, and environment 82–93

Wallace, Alfred Russel 28
Walpole, Horace 85
war films 112
Waste Land, The (Eliot) 103–5
Weltanschauung (lifeworld) 168 (n42)
western world: capitalism 44–8;
dichotomies 162; literate
constructions 11–13
White, Lynn 129

Whitman, Walt 105
wilderness 14, 15, 112
Wittgenstein, Ludwig 19
women 143, 144–7
Women in Love (Lawrence) 97–8
Wordsworth, William 102

world-view: transmuted 57; women and
  nature 145
Wright, Frank Lloyd 115

Zen Buddhism 106, 132, 148–9
Zoroastrianism 158